BARBARA LOBO

ENVIRONMENTAL CRISIS:

FLEEING FROM CHAOS

ISBN: 9798476345893

Editorial and research content: **Barbara Lobo**
English language correction/ editorial review: **Sharon J. Bowers**
Cover image: Canva/ Barbara Lobo

Environmental Crisis: Fleeing From Chaos
LOBO, Barbara

1ª Edition
2021

I dedicate this book to everyone who want a better, preserved, diverse and happy world.

"Earth provides enough to satisfy every man's needs, but not every man's greed."

Mahatma Gandhi

CONTENTS

PREFACE

First, I would like to say how privileged I feel to be among the first people to access this very essential work prepared by Dr. Barbara Lobo. Then, I would like to record how much I feel honored and grateful to write the preface to this book. I have been following Dr. Barbara Lobo's work for some years now, as a writer, researcher and professor at the undergraduate and graduate levels in the field of technology, and along this journey my admiration for her work has been growing.

Dr. Barbara carries a sensitivity as a researcher to produce knowledge of the highest relevance, aligned with environmental and social issues. Such knowledge directs us to possible paths for the design of a better society for all, therefore, surely, with this publication, "Environmental Crisis: Fleeing from Chaos" was not different.

The book brings to the reader a discussion about the various laws that aim to deliberate on environmental issues, considering the particularities of different countries, especially aiming the protection and prevention of biodiversity. It also brings an important reflection on the relationship between urban growth, consumption, and environmental issues, especially when positioning us against a modern society that faces problems related to excessive consumption. On the other hand, there is the under-consumption, resulting in standards of quality of life so unbalanced globally.

In this sense, it will be observed that disordered growth not only strengthens the problems that are already in place but also maximizes the consequences for the environment and the alternatives to ensure a better future. Therefore, given the importance of this agenda, the book intends to share the different contexts in which this discussion has been expanded.

The author also draws the attention of managers on local, national, and international levels, about the problems of urban planning, waste generation, and pollution, among others. Waste generation continues to be a global problem and numbers confirm the importance of developing strategies and public policies to minimize this global bottleneck.

The problem of waste generation is aggravated, it originates from the various channels, including urban and agricultural resources and industrial waste, which are constantly dumped into the air, water, and soil, making it increasingly difficult to absorb this increasing volume of waste. In this sense, pollution has been causing damage to the ecosystem as a whole, including the land, fauna, flora, and human health. Therefore, it has an impact on biodiversity and the quality of life levels of the global population.

In this publication, Dr. Barbara shows the influence of urban pollution on the system, however, it also indicates how large cities have been increasing the problem in ever greater dimensions, given the emission of carbon dioxide (CO_2). It is known that CO_2 is one of the main causes for the greenhouse effect and contributes especially to climate change, in which among the main drivers are the gases generated by motor vehicles.

The book presents important data about deforestation, a problem with great global impact, responsible for the extinction of land animals and for putting a significant percentage of animals at risk. Relating also as a result of degradation, direct impacts on biodiversity, society and favoring the increase of public health problems. In this context, the author highlights the importance of the Amazon, as well as the union between the various stakeholders to ensure the conservation of this forest, considered the largest tropical forest in the world, which unfortunately has been the target of deforestation and fires.

The facts reported so far, without being enough for the installed environmental crisis, the book will show a new milestone, where the world is facing a COVID-19 pandemic, which generated global impacts on the entire society, where problems were aggravated with the volume of deaths, diseases, unemployment, among other problems that came with the pandemic, requiring, therefore, new standards of living and a new normal. In this sense, discussions about sustainability, or sustainable development, were based, as the public authorities in many places have difficulties in complying with legislation, executing sustainability projects, and identifying alternatives between economic growth and the conservation of natural resources.

Around a broad discussion about the Cartesian paradigm, Dr. Barbara advocates for an ethics of responsibility associated with the ethics of compassion, envisioning a new look at nature, as well as an alternative to individuals, aiming especially at the balance between reason, compassion, and responsibility. In chapter five, a historical journey is presented about the agreements, researches, and initiatives between countries, which were driven around 1970, by the volume of accidents and environmental problems installed.

This journey reaches the year 2020, in which the Sustainable Development Goals Report and the 2030 Agenda for Sustainability were developed. It was also considered the Brundtland Report, which argues that poverty is the main cause of environmental degradation and points to a tendency towards Ecocapitalism. Given this, the book shows that sustainability remains an economic challenge for the coming decades.

And finally, Dr. Barbara opens an important discussion for the new environmental age, and among the possibilities to get rid of Chaos are sustainability partners. For that, the axes of environmental education, recycling, and population participation will be considered. And the author will brilliantly defend the need to think about strategies based on participatory processes, among the various stakeholders, including the neediest layers of the population. Therefore, environmental education must have the mission to help of ensuring a healthy environment for the population as a whole and the types of life existing on the face of the Earth, stimulating behavior change. And about recycling, it is highlighted that although it is fundamental, it is also advocating for the reduction of consumption and intends to encourage the engagement of people and organizations with discussions regarding environmental problems.

Certainly, as you read, you will notice special care with the historical journey of the themes that this book intends to address, you will identify important data at a global level that contributes to a quick visualization of the growing problems towards Chaos as well as a highly relevant conceptual discussion that will contribute to the reflective process on the environmental crisis and the possible ways to impact part of the population through education, consumption reduction, and engagement in environmental issues.

Therefore, I recommend a sequential reading of the chapters, taking some time to reflect on the installed problems and still seeking to understand your role in the planet's transformation process. In the face of Chaos, collective actions are essential, aiming to meet the need for protection and conservation of biodiversity, the better quality of life indexes for humanity, and organizations with systems of products and services largely based on sustainability and engaged with collective objectives.

Prof. Thiago Gomes de Lima
Ph.D. in Environmental Sciences and Conservation from the Federal University of Rio de Janeiro (UFRJ), with a Sandwich Ph.D. in Management Engineering from the Technical University of Denmark (DTU). He is currently a Professor at the Superior Magisterium at the Polytechnic Institute of the Federal University of Rio de Janeiro – Macaé, Brazil.

ACKNOWLEDGEMENTS

I would like to thank Mrs. Sharon J. Bowers for accepting my proposal for this book in collaborating with editorial review and to be part of my professional life helping me to reach a better English each day more.

I am also grateful to my dear friend and university professor at Estacio University where I have worked for almost fifteen years in Brazil. Professor Doctor Tiago was my Supervising Director and choose me to be a coordinator of postgraduate course in Petroleum Engineering at Estacio, and since then we have become friends. Thanks to him to accept my invitation for writing the preface of this book and for doing it so grandly.

Specially, I owe huge thanks to my husband. Alberto is a wonderful person to me, and he has been patient to kind and supporting my writer path ever.

Thanks my children, Luiza for giving me all the love I need and Gabriel for your life.

Finally, thanks my mom for supporting me since I was born.

And above all, thanks God. Lord, you are good all the time!

INTRODUCTION

The close relationship between health and environmental issues speaks to the development of this book, because climate crises, pollution, widespread disease and environmental imbalance affect social dynamics, reducing the quality of life and causing the extinction of species.

Maintenance of the environment is one of the necessary conditions for quality of life. It is necessary to plan urban space in order to ensure the conservation and control of the use of natural and artificial resources, waste management, technical, acoustic, visual and spatial comfort. That is, environmental conditions must be created that reduce or avoid the risk of exposure of the population of health problems.[1]

There is a need to expand public engagement through efforts to raise the level of awareness of all residents of planet Earth. It is vital

---

[1] PHILIPPI JR., 1988.

to stimulate popular participation in the process of environmental preservation and coping with the climate crisis, which has affected all humanity and caused environmental imbalance and the destruction of the planet.

With an eye on the environmental, the climate crisis and the increase in industrialization, production and consumption, we perceived the need to discuss sustainable development, environmental education, and actions to protect and preserve nature as allies to the balance that the environment requires.

An important objective of this book is to identify and demonstrate some of the most urgent global environmental problems. After a brief history or the global perception of this crisis, the discussion was expanded to address the problem through environmental education, popular participation, governmental responsibility and scientific credibility.

Scientists have developed research and metrics that measure important the environmental markers, problems, and its confrontation. May this truth be heard by all of us!

The first chapter *The Environment and Its Laws* discusses the development of concept of the environment, its implications for society, and its framework of laws.

The second chapter *Environmental Crisis* presents the environmental crisis with metrics on urban growth and consumption. This chapter shows an analysis of waste generation, climate change and deforestation as problems have contributed to large-scale environmental crises.

Pollution data is also presented and demonstrates that greenhouse gases act as accelerators of global warming and climate change.

In chapter three *Sustainability* the theme presented is the nuanced definitions sustainable development and sustainability policies.

The fourth chapter *Cartesian Paradigm and Its Influence* shows the effect of the Cartesian method of man's relationship to the environment. Individualistic and uninterconnected actions and thoughts present themselves as one of the causes of environmental imbalance. This demonstrates that in only a few centuries man has advanced economic and simultaneously "development" put himself in danger and destroyed several species.

In chapter 5 *Government, Summits, Agreements and Scientific Research: The B Side of History* presents a historical view: dilemmas and criticism of the environmental debate here, we emphasize the real need to create strategies, at all levels and in particular politics for the construction of a conscious and participatory society.

By pointing out the dilemmas of sustainable development, Herculano (1992) presents the historical antagonism between the concepts of sustainability and development.

The sixth chapter *The New Environmental Era – New Rules to Fleeing from Chaos: Sustainability Partners* emphatically discusses environmental education, recycling, popular participation and science as important allies in coping with the environmental crisis.

This chapter discusses a variety of ways to increase social participation in relation to environmental problems: creation of a cohesive collective consciousness, and ultimately, united mobilization.

Educational policies also emerge as instruments for critical reflection where education becomes allied to human development and an important instrument of social transformation.

Environmental education is relevant in the context of building communal concepts of sustainable development, emphasizing the importance of popular participation in the development of this process. Popular participation is absolutely vital. Individuals must find through engagement. Only then can we achieve a planet that finds and maintains equilibrium.

And finally, science is explored as the holder of the truth about the environment. It must be understood as the solid foundation to break through the chaotic barriers of misinformation, dangerous rhetoric and propaganda.

The administration of the environmental crisis is a scientific, a technical, a political and an environmental education issue and these are a problem for all of us.

1

THE ENVIRONMENT AND ITS LAWS

"Every time fences fall, society is forced to look at itself and discuss the size of inequalities, the size of opulence and misery, the size of abundance and hunger". (Pedro Tierra, poet, 1995).

This chapter discusses environmental conditions, how the environment is defined, its historical evolution, legal aspects and dilemmas in achieving sustainable development processes and policies.

The environmental first consolidated as an international issue in the Stockholm Declaration of 1972, enabling a global look at the environment. Protection of the environment, with an emphasis on sustainable development, arose from the First United Nations Conference on Environment and Development in Sweden. At this meeting, 26 fundamental principles were established as international prerogatives focused on the environment.

When analyzing the concept of "the environment", it is necessary to understand that its protection is articulated in a multidimensional way. Environmental rights are protected through international regulatory systems, as well as national and local systems; these vary according to specific local environmental issues as well as regional problems.

The definition of "environment" may vary according to the source of observation. Generally, environment can be understood as "a compound of biotic and abiotic beings distributed in physical, chemical, biological and social elements and their interrelationships, endowed with the ability to integrate natural systems and cause effects on them." In other words, wherever there are living beings, and their interaction with the environment composed of other living beings, organic and inorganic matter. For the United Nations (UN) the environment is "the set of physical, chemical, biological and social elements that can cause direct or indirect effects on living beings and human activities." According to the Brazilian Ministry of the Environment, the same translates into "the interaction of the set of natural, artificial and cultural elements that provide the balanced development of life in all its forms."

Due to the environment requiring protection through legal documents, norms, national and international laws, political overall, it has also become an important elective matter in the constitutions of several countries. Environmental law is a complex web of legal practices oriented to facing a variety of global and regional environmental problems.

Environmental law concerns law relating to environmental troubles. We can compare this with many anthropocentric actions including philosophical, political, and societal questions. Fisher includes relative impacts of a collective nature, such as deforestation, species extinction, soil, air and water pollution, anthropocentric climate change and so on.[2]

It is important to note that environmental law includes the deliberation mechanisms of legal systems capable of addressing environmental issues that are complex and biodiverse. In international law, signing environmental agreements between nation-states is inconsistently reinforced. Complicating matters significantly signing agreements aimed at global environmental problems often leads to fragmentation of international law. International Law Commission has said that fragmentation results in conflicts between rules and divergent institutional practices, causing the loss of a unified understanding of those laws.[3]

Each country makes its environmental law and rules for the protection and preservation of biodiversity. When the environmental issue involves a large scale problem, international agreements are created as a solution. These agreements carry a

---

[2] FISHER, 2017.
[3] FISHER, 2017.

system for monitoring their actions as a guarantee of their practical applicability. There is usually a creation of committees or systems that establish plans, goals, and actions to face the local and global problem. These regulatory systems are widely discussed at international meetings, summits, and conferences.

Protecting the environment as a fundamental right in its national constitution is a way to preserve environmental resources broadly. In some countries, such as Brazil and Portugal, the Constitution provides "a fundamental right to a healthy environment." However, in others, such as in the United States, it is established in the National Environmental Policy Law (NEPA). In Germany environmental protection is referenced in the constitutional norms.

The Brazilian Federal Constitution of 1988 dedicated a chapter to the environment, as article 225:

> Everyone has the right to the ecologically balanced environment, well of common use of the people and essential to the healthy quality of life, imposing on the Public Power and the

> collectivity the duty to defend it and preserve it for the present and future generations.

The Portuguese Constitution (1976), in Article 66, provides this article on environmental matters under the title Environment and Quality of Life:

> 1. Everyone has the right to an environment of human life, healthy and ecologically balanced and the duty to defend it.
> 2. In order to ensure the right to the environment, in the framework of sustainable development, it is incumbent on the State, through its own institutions and with the involvement and participation of citizens.

In the United States, environmental legal protection is at the discretion of the National Environmental Policy Act (NEPA, 1969). NEPA more recently changed (42 U.S.C. 4321-4347) requirements for federal government agencies and established the Council on Environmental Quality. Referring Section 101(c), NEPA states that it is the right of citizens to enjoy a healthy environment, and assigns responsibility to conserve of the environment. According to the law:

"Among the most significant features of the law in Section 102(2) are the requirements for federal agencies to use a systematic, interdisciplinary approach to ensure integrated use of environmental arts in planning and decision making which may have an impact on the human environment; to develop procedures to ensure that environmental amenities and values are given appropriate consideration in decision making, along with economic and technical considerations; and to include in every recommendation or report on proposals for legislation and other major Federal actions significantly affecting the quality of the human environment, a detailed statement by the responsible official." (MARRIOTT, 1997).

The German state constitutes a Federative Republic in accordance with Bonn Fundamental Law (1949). Legislative powers are shared by the Union (Bund) and the States (Länder), yet there are no specific details that propose environmental protection as a fundamental right. Since 1970, Germany has enacted a coherent environmental policy, welcomed by the Bonn Fundamental Law through constitutional revision (1994), article 20A4.

The Brazilian legal framework is nicely aligned with the principles of sustainable development, encompassing a set of constant public policies: National Environmental Policy (1981), Brazilian Constitution (1988), National Water Resources Policy (1997), National Environmental Education Policy (1999), Environmental Crimes Act (1999) and City Statute (2000) and other more recent ones. Added to this set is an immense list of laws at the state and municipal levels, favoring the development of environmental control actions, the responsibility of polluting agents and prioritization of environmental education.[4]

In Brazil, environmental protection is supported by the Federal Constitution, Title VIII, Chapter VI, in non constitutional laws, highlighting, among others, Law 9.605/1998. These provide criminal and administrative sanctions derived from conduct and activities harmful to the environment, and may penalize, through environmental agencies, legal entities in the case of observance of crimes against the environment.

---

[4] MALHEIROS, 2005.

As set out above, there are several other laws dealing with environmental matters in Brazil.

For example, Law 6.938/1981 aims to preserve, improve and recover environmental quality conducive to life. This law established the National Environment Policy and the environmental protection system, "SISNAMA", as a normative instrument that encompasses various themes within environmental law. This document regulates the proper use of environmental resources and the guarantee of environmental licensing, obliging the polluter to indemnify for environmental damage caused, regardless of fault.

Law 9.433/1997 created the National Water Resources System, and defines water as a limited natural resource, endowed with economic value, which can have multiple uses: human consumption, energy production, transportation, sewage release. Thus, because it is a limited resource and endowed with economic value, this law instituted regulations for the collection of water, conditioning the intervention in public waters to the authority of a competent institution.

In Brazil, forests have their protection and preservation guaranteed through the New Brazilian Forest Code, under Law 12.651/2012, which provides the protection of native vegetation. This law guarantees that forests existing in national territory and other forms of native vegetation are recognized as goods of common interest to all inhabitants of the country. Moreover, the right to property is conditional on exercise in the form of the law. This law replaced the Brazilian Forest Code of 1965.

Law 6.902/1981 on Environmental Protection Area is an important law for protecting areas that have representative ecosystems in Brazil. The law establishes guidelines regarding the creation of Ecological Stations, which are areas composed of different ecosystems that need to be preserved, and Environmental Protection Areas, which are areas regulated by the government; these balance interaction between economic activities and environmental protection.

Another relevant law is Law 12.305/2010 known as the National Solid Waste Policy. This policy establishes principles, objectives, instruments and guidelines for integrated management of solid waste, including hazardous ones, and defines the responsibilities of generators and the government. The standard also establishes the need to prepare the Solid Waste Management Plan, for future policies.

Law 6.803/1980 deals with industrial activities and zoning. It provides basic guidelines for industrial zoning in critical pollution areas in Brazilian territory. Also it establishes environmental standards for the installation and referred licensing of industries, where an Environmental Impact Study is required.

It is important to highlight the laws, norms, policies and all regulatory framework of environmental law as fundamental sources of protection and preservation of the environment and also to define the criteria of guilt and legal responsibility for actions harmful to natural resources. These instruments hopefully function as disincentives to polluting and conditioning public the use of available natural resources, to be harmonious with sustainable development.

Environmental law is a systematizing right, which articulates legislation, doctrine and jurisprudence concerning the elements that integrate the environment.[5] In order to analyze environmental law, it is important to observe the relationship between economic development and environmental protection and its collective interest in the two assumptions.

> The economic nature of environmental law cannot be noted as a type of legal relationship that privileges productive activity to the detriment of a minimum standard of living that must be ensured for human beings.[6]

Man has a fundamental right to satisfactory living conditions; these include a healthy environment that allows him to live with dignity and well-being, in harmony with nature and education about defending and respecting these values. Thus, popular

---

[5] MACHADO, 2002, apud ANDRADE, 2017.
[6] CAVALCANTI, 2004, apud CARDOSO, 2016.

participation and information are actions of extreme relevance to develop a critical awareness for the development of these concepts and actions. The biggest problem is social exclusion and income inequality. This disparity a lack of governability and public policies of social inclusion. Or is it unethical?

When we speak about social development and environmental rights, we need to think about various social exclusions that include ethical and moral dimensions. Social exclusion results when government and social institutions do not work together to reach better human development. The economic impact is the main cause of non-investment in social politics. If government raises taxes on industries, it will lead to economic development at lower growth rates. The impact is both economic and social. There is a great for consistent need investment and maintenance policies as well as social programs.

Social ethic education should be connected to social rights that are guaranteed by democracy. Social ethics reveal how we can deal with avoiding group conflicts. Personal ethics tell us a little bit about our character, our individual morals built on our vices and virtues. The main idea is that all governments should answer these questions: What is your mission? Why am I here at all? Whose got what rights? Whose got which responsibilities?

The change in human actions implies a change in ethics. To conceive a new action that does not include the environment as worthy of rights and duties is to remain stuck in the Cartesian.[7]

For the above author, the desecration of nature and civilization develop together. Both rebel against the elements. The first, by penetrating them and raping their creatures; and the second in the fallout from the resultant collapse of social systems and civic life.

It is imperative not only to preserve, but to establish harmony between nature and man, to ensure an ecologically balanced and healthy environment.

---

[7] JONAS, 1995.

The great need for balance and preservation of the environment is to emphasize that meeting the basic needs of the entire current and future human contingent will increasingly demand the intelligent use of environmental resources, and the modification of most ecosystems. This requires us to consider water, soil and air not only as components of the biosphere, but as resources that can and should be exploited, respecting their ability to sustain human life and culture.[8]

In conclusion, comprehensive and cooperation studies and research on environmental matters, the environment conceptually began to come into focus on a planetary scale from the beginning of the 1970s. Mass media has perpetuated criticism about borders created to separate the human world from the natural world. The fourth estate most definitely has a role to play to continue our efforts to conserve, preserve, and wisely steward our shared world.

---

[8] MUCCI, 2005.

Nature does not follow the order instituted by humans and its time is not measured by the clock. Man does not hold time, but can slow down the entropic and evolutionary process, favoring the future of our species and the flourishing of our planet.[9]

---

[9] TIEZZI, apud LA TORRE, 1993

2

ENVIRONMENTAL CRISIS

2.1 Urban growth, consumption and environmental issues

The environmental issue involves all sorts of problems regarding the socio-environmental conditions of urban and non-urban areas. Their definition includes many aspects of the quality of human life, the impacts of human action on climatic[10], hydrological[11], geomorphological[12], pedological[13] and biogeographic[14] conditions, on all time and space scales.[15]

Climate conditions affect environmental dynamics, influence society, the economy and the use

---

10 Climate change: Female climate plural; adj. on climate: climate influence.
11 Hydrology: Science that deals with waters, their properties, laws, phenomena, interaction with the environment and distribution, on the surface and below the earth's surface.
12 Geomorphology: Study of forms, characteristics and processes related to terrestrial relief; geomorphy, geomorphenia, morphology.
13 Pedology: Science that studies the physical, chemical and biological characters of soils.
14 Biogeography: Branch of biology that deals with the geographical distribution of different species of animal and plant living organisms and the factors that influence this distribution.
15 SOBRAL and SILVA, 1989.

of natural resources. They are also responsible for processes such as constitution, structure, development and distribution of natural resources, including soil, relief, water resources, fauna and flora.[16]

The environmental degradation that has occurred over time is due to the profound social, economic, philosophical and political changes that have reached all humanity. We have introduced values and unwittingly practices that are at odds with the basic necessities to maintain a healthy environment and allows all members of society a high quality of life.[17] This degradation has been increasing and its impacts are now severe, especially with regard to climate change. In 2021 environmental matters have been a constant concern to the scientific community due to a false narrative of the far right. This trope distorts the threat to industries, to individual citizens. It disparages the government's

---

[16] SANTOS, 2000.
[17] PELICIONE, 2005.

responsibility to protect everyone from environmental damage.

In analyzing society, we observed that our way of being, producing and living is the result of a way of thinking and acting in relation to nature and other human beings that have been outdated for many centuries. This dichotomy between human beings and nature has been misunderstood since ancient Greece, thus demonstrating the non-integration of man/environment.[18]

The relationship between modern society, including social problems, and environmental issues underlie a remarkable dynamic in the historical context to which society is inserted. It is pressing to comprehend the social context when we seek to question environmental issues in order to relate social and environmental policies and guarantee individual rights.

---

[18] GONÇALVES, 1990.

When discussing the environment, it is necessary to take into account sweeping historical changes in consumption and production patterns.

The consumption pattern of a society can be defined by the quality and quantity of use of natural resources for the production of consumer goods and how effectively this meets society's demand for food, housing, transportation, leisure, etc, it a ratio of exploitation and transformation of natural resources to meet human needs.[19]

It is well-known that current consumption patterns demonstrate problems related to over-consumption and underconsumption. The implied contradiction presents growing consumption on a global scale and in parallel millions of people not consuming enough to meet their basic needs. These two analyses cause great stress to the global environment (UNEP). According to Laslo (2004), 16 percent of the world population uses 86 percent of consumer goods, while 84 percent of the world

---

[19] PHILIPPI JR. and MALHEIROS, 2005.

population survives on just 14 percent of available goods.[20]

When we discuss the environment from the point of view of consumption, we reiterate the concern with non-renewable environmental resources, the absence of effective environmental protection and preservation actions and pollution as a negative impact on the environment. It is crucial to reinforce that the process of pollution and the depletion or scarcity of natural resources contribute to the loss of biodiversity, impacting environmental quality and affecting the planet's capacity to absorb waste that causes further degradation of renewable resources.

The human community needs more balanced living standards, across the board: a good quality of life, as defined by the economic, social and human development. From the perspective of quality, it seems obvious that the relationship between man and nature must be better balanced: decent health conditions, efficient sanitation and environmental

---

[20] RODRIGUES, 2007.

balance. Pollution is an inhibitor of human, social and environmental development when it comes to sustainability.

According to the 2030 Agenda, on sustainable development goals, sustainable production and consumption patterns, it is pressing:

> Change in consumption and production patterns as indispensable measures in reducing the ecological footprint on the environment. These measures are the basis of sustainable economic and social development. The goals of SDG12[21] aim to promote the efficiency of the use of energy and natural resources, sustainable infrastructure, access to basic services. In addition, the objective prioritizes information, coordinated management, transparency and accountability of consumer actors of natural resources as key tools for achieving more sustainable patterns of production and consumption.

We can gain a fuller understanding of the relationship between sustainable development and economic growth if we compare human growth dynamics with the biodiversity and Earth's

---

[21] SDG12 – Sustainable Delvelopment Goals (SDGs), aiming at achieving the targets set in 2015, according to SDG 12, which deals with ensuring the sustainable production and consumption standards established by the UN - United Nations.

sustainability. It can appear contradictory, but according to Sachs (2015) we need to learn how to achieve economic growth that remains within planetary boundaries. It is not an easy feat.

Sachs (2015) points out how pent-up growth presents a challenge to the world economy. It means the amount of economic growth necessary for poorer countries to reach the richer countries in the future. That is, assuming the richest countries do not grow faster than others.

We can understand this by looking to convergence theory. This theory shows poor countries must have high growth rates because they begin with the challenge to overcome poverty. According to Sachs (2015), the poorer a country's starting point (of course, no poverty trap or another barrier to growth), the greater the headroom for fast catching up. Therefore, this theory explains how poor countries achieve faster growth and move closer to richer countries. Convergence theory reveals that as they become closer technologically and economically their growth rates also slow down.

By the way, if this economic model is achieved with new technologies, over time the world will overtake its planetary boundaries and then we will have more environmental problems because of consumerism and industrialization process among other things. In other words, Sachs (2015) warns: If we want the world economy to develop in a fundamentally different way in the future, we need to reconcile this growth with the ecological realities of planet Earth.

In other words, all these issues are the reflection of unequal development. We know that the planet already has exceeded its biological capacity to absorb the waste it produces. This overload has been felt for years with accelerating environmental impacts.

Furthermore, to understand the complexity of their consumption model it is necessary to address the increase in population. The world population growth in the last century, when associated with the increase in the consumption rate of natural resources and the accelerated process of urbanization, resulted in unprecedented urban pollution.[22]

According to United Nations Department of Economic and Social Affairs (2019) shown in figure 1º we can see how population growth has evolved in the last fifty years. In 1950 the world population was around 2,536,430,000 and currently, in 2020, the number of people is 7,794,799,000. This analysis presents an estimated world population in year of 2050 around 9,735,000,000 people. In 2100 the world's population is expected to be more than 10,875,300,000 people.

According to this global demographic outlook, the world's population is expected to grow by 2 billion people over the next 30 years. That is, the estimated population growth is 7.7 billion individuals in 2020 to 9.7 billion in 2050.[23] By the end of the twenty-first century the world will reach almost 11 billion people.[24]

---

[22] MALHEIROS, 2005.
[23] UN, 2019.
[24] SACHS, 2015.

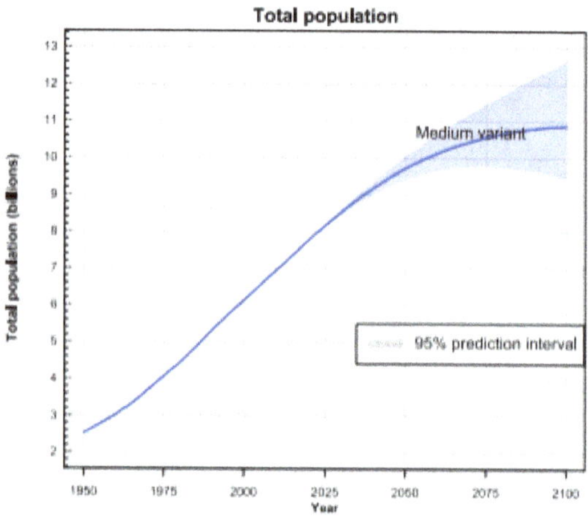

1 World population prospects 1950-2100

Source: United Nations Department of Economic and Social Affairs Population Division (DESA Population Division). 2019. "World Population Prospects 2019." New York. Adapted by author.

We can see now that the planet is literally exploding. If consumerism like grows population and without regard for planetary boundaries we will be in collapse. We must discover new technologies and economic models to remain viable in the future.

This will mean we need the world to develop in a completely different way in the future if we want to reconcile the growth with the ecological realities on the Earth.[25]

Growing and disorderly urbanization and its consequences for natural communities have always been part of the agendas of different international meetings on environment and sustainability. To name a few: the Club of Rome, Stockholm Conference (1972), the Brundtland Commission (1983), Rio 92 in Rio de Janeiro, Brazil (1992), Rio + 10 in Johannesburg, South Africa (2002), Rio+20 in Rio de Janeiro, Brazil (2012), and finally, the Sustainable Development Summit in New York, United States, at the headquarters of the UN (2015).

Despite the urgency of the numbers, cities continue to grow around the world, consuming resources and contributing to inexcusable loss of environmental quality, especially related to water resources.[26] Within this scenario, the education and awareness of present and future generations has an indisputable value in the process of change of attitude,

---

[25] SACHS, 2015.
[26] GRANDISOLI, 2005.

creating new solutions to the problems brought about by the lifestyle of the population.

Thus, with the advent of the dense occupation of cities in the period of modernity, major changes have been suffered by the environment, pollution, the production and dumping of natural and artificial waste in rivers; this is only example of urban environmental degradation.

We see how economic and technological development, population growth and consumerism have caused significant adverse effects on environment. In 2017, a paper published in The Anthropocene Review estimated the cumulative total of material output of collective human enterprise since the Industrial Revolution at 30 trillion tons.[27]

The degradation caused by the occupation of large cities, from the Industrial Revolution to modernity has resulted in scientific, political, legal and social concern. Long and increasing international and national debate has reflected a particular moment

---

[27] ZALASIEWCZ et all, 2017, apud O'NEILL, 2019.

in the relationship of Western man with nature, with the economy and, with other men, as well as of his mastery of the techniques and human uses of natural resources. We can no longer deny that the main concern has been with the impact on the environment since the beginning of urban, industrial and technological development.

2.2 Waste generation, pollution and environmental issues

The environmental problem has been featured in several national and international forums; the idea that environmental degradation is a consequence of human actions and now, requires action on the part of all governments and societies.[28]

It is important to note that governmental management can make a positive difference in terms of quality of life, social development, economic, among others. On the other hand, the inefficiency of

---

[28] MACHADO, 2003.

governments in the planning and execution of public policies causes problems for humanity.

When issues of collective interest, such as urban development and waste, are not observed, the result is environmental pollution and negative health impacts.

Around the world governments, companies and everyday people have a collective responsibility to prevent adverse environmental impacts on life's quality. This means that we need to make informed decisions daily about production and consumption. When public policies are not founded on good management, the quality of life diminishes. Decision-makers must consider many data points: urban planning, population growth and waste generation, public hygiene, garbage collection, disease transmission and increased pollution. In particular, waste generation and disposal is a big issue that must be understood by all levels of decision-makers from local to regional, national, and international.

According to the World Bank, we generate an amount of solid waste that is demonstrably unsustainable. Annually we generate 2.01 billion tons of solid waste just in cities.

The waste generated per person averages 0.74 kilogram, however, this range varies widely between 0.11 to 4.54 kilograms. The richer countries generate around 683 million tons of the world's waste, that is 34 percent.[29] According to UN, this production could reach in 2.2 billion tons a year within ten years. In the middle of this century with an increase in population to 9 billion this production projects 4 billion tons of urban waste per year. The World Bank (2018) estimates global waste will probably increase to 3.40 billion tons by 2050.

According to Federal Senate studies from Brazil, the Nations belonging the Organization for Economic Cooperation and Development (OECD) are responsible for consumption more than 60 percent about all industrial raw materials.

There is a natural relationship between urban growth and waste generation. Economic

---

[29] WORLD BANK, 2018.

development and urbanization processes walk together, and both have contributed to a steep increase in pollution throughout the history of humanity.

In research conducted by the World Bank (2016) we perceive an elevated rise in waste production compared to years before. That is, in 2012 global waste production was 1.3 billion tons a year. In 2016 this estimated number rose to 2.01 billion tons a year.[30]

The countries of Europe and Central Asia and East Asia and Pacific regions generate most of the world's waste, around of 43 percent a year on average. The least amount of world's waste was produced by countries of Middle East and North Africa and Sub-Saharan Africa, around 15 percent a year of the world's waste.[31]

---

[30] WORLD BANK, 2018.
[31] WORLD BANK, 2018.

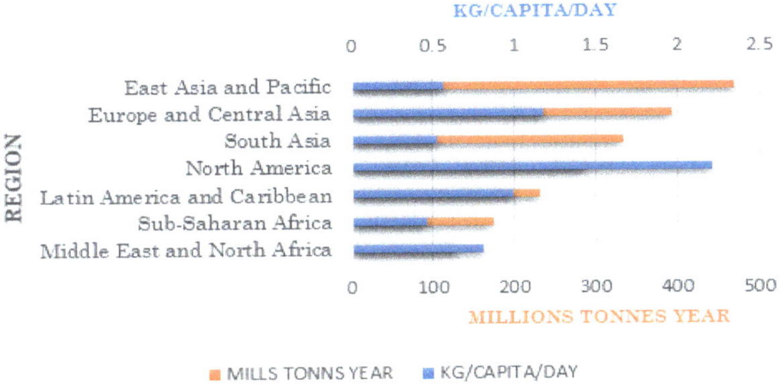

2 Waste Generation by Region

Source: World Bank, 2018. Created by author, 2020.

According to Figure 2, we can see the waste produced by region. And also the comparative per capita figures. The east Asia and Pacific regions produced the most annual waste. When we compare this production in kilogram per capita a day, this region was the third among the others with the least waste generated. That is, second only to South Asia and sub-Saharan Africa; this region generated the least per capita waste per day in 2016.

To understand more about world's waste generation rates we need to analyze each region and their respective countries based on the World Bank studies the year 2016. We can know this better from the bellow:

We start with the East Asia and Pacific Region and Europe and Central Asia because both regions are responsible for 43 percent of the world's waste.[32]

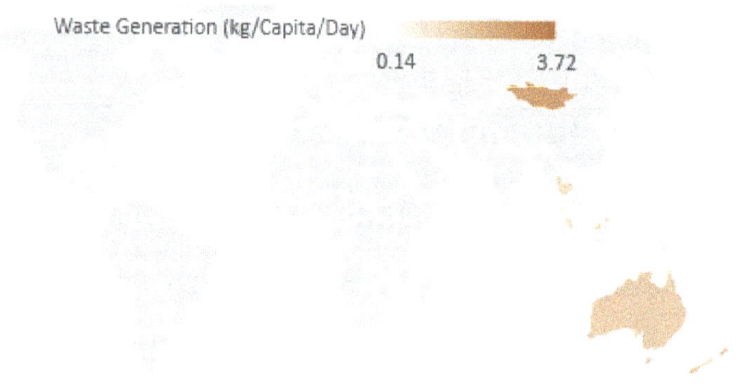

3 EAST ASIA and PACIFIC REGION

Source: World Bank, 2018. DSAT Editor, 2020. Created by author, 2020.

The East Asia and Pacific region, in figure 3, has 37 countries and they generated 486 million tons of waste in 2016. Their per capita average was 0.56 kilogram a day.

---

[32] WORLD BANK, 2018.

Singapore, Mongolia, Guam, Hong Kong - China, New Zealand, Macao SAR - China, Northern Mariana Islands and Australia, lead the rank of waste generation rates in this region, between 1.54 and 3.72 kilogram per capita day. China produces about 47 percent of the region's waste. Although this is an elevated rate, both China and Republic of Korea have recently been invested in landfilling and recycling.[33]

4 EUROPE and CENTRAL ASIA REGION

Source: World Bank, 2018. DSAT Editor, 2020. Created by author, 2020.

Europe and Central Asia (figure 4) produced 392 million tons of waste in 2016, that is,

---

[33] WORLD BANK, 2018.

their generation per capita a day was 1.18 kilograms. This region includes 57 countries.

At this time around 31 percent of waste is recycled and/or composted. Western Europe is responsible for higher recycling and collection rates. Eastern Europe and Central Asia are looking to modernize their waste management systems. By the way, waste prevention and recycling are rising in all regions. Good news indeed!

Iceland, Faroe Islands, Monaco, Moldova, Channel Islands, Greenland, Liechtenstein, Denmark, Isle of Man, Ireland, Germany and Luxembourg had the highest waste generation in kilogram per capita day. They produced rates between 1.72 to 4.45 kilogram per capita a day.[34]

---

[34] WORLD BANK, 2018.

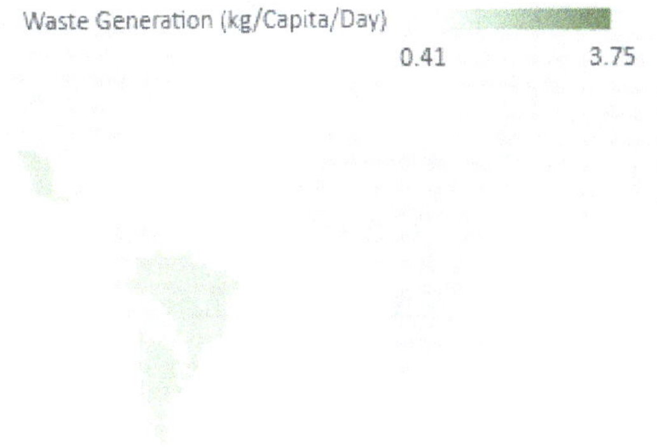

5 LATIN AMERICA and CARIBBEAN REGION

Source: World Bank, 2018. DSAT Editor, 2020. Created by author, 2020.

The region of Latin America and the Caribbean (figure 5) includes 42 countries and South America is part of this. In 2016, this region generated 231million ton of waste, and represents 0.99 kilogram per capita per day on average.

Excluding the Caribbean Islands, the recycling is common in the urban areas, mainly material such as paper, plastic, and aluminum.

By the way, from a policy vision standpoint, most countries in this region have at least one regulatory system to conduct waste management actions.

US Virgin Islands (USVI), British Virgin Islands, Cayman Island, Puerto Rico, Aruba, Bahamas, Barbados, St. Martin (French part), St. Kitts and Nevis, Trinidad and Tobago, St. Lucia, Mexico, Chile, Argentina, Dominican Republic, Brazil, Panama and Uruguay are the ones that produce the most waste in kilogram per capita a day. They range from 1.01 to 4.46 kilogram per capita day.[35]

In Brazil, until the mid-1990s, the population produced and released daily into the environment about 240,000 tons of solid waste, of which an insignificant portion had health consequences.[36]

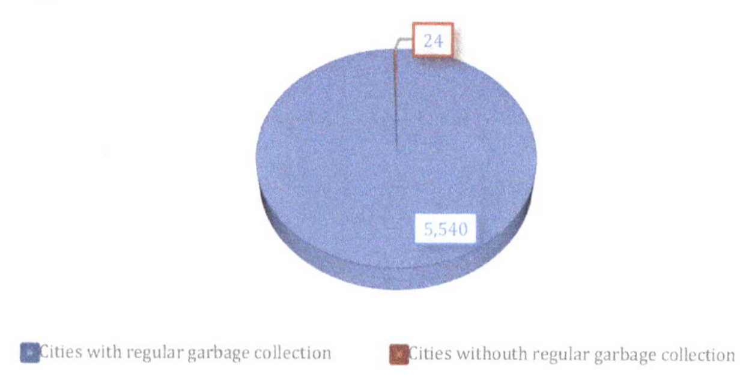

6 Brazilian Cities: Solid Waste Collection

Source: Brazilian Institute of Geography and Statistics (IBGE), 2008. Created by author, 2020.

[35] WORLD BANK, 2018.
[36] MACHADO, 2003.

A survey conducted at IBGE[37] in 2008, aimed to determine the reality of basic sanitation in Brazil; provision of solid waste management services; the quality of those services; and the conditions that directly influence the environment.

IBGE concluded (figure 6) that in a total of 5,564 municipalities, of the 5,565 existing in that year, 5,540 had regular garbage collection. This represents 99.57 percent.

The selective collection of recyclable solid waste comprises 17.85 percent. The screening of recyclable solid waste is 17.79 percent. On the other hand, the collection of demolition and construction waste is higher than the previous ones, comprising 71.62 percent. On the collection of special solid waste, which includes health and industrial waste, the value reached is 80.23 percent.

Regarding the treatment of solid waste, the lowest value was seen for this type of service among the others, with a total of 16.82 percent.

---

[37] Brazilian Institute of Geography and Statistics – IBGE.

Finally, on the disposal of solid waste in the soil, a very impressive value of 82.39 percent is shown, behind only the total number of regular garbage collection, see Figure 7.

7 Cities with solid waste management services

Source: Brazilian Institute of Geography and Statistics (IBGE). 2008. Adapted by author, 2020.

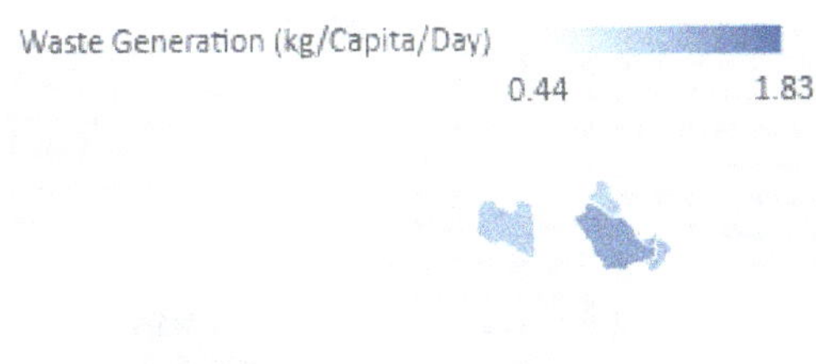

8 MIDDLE EAST and NORTH AFRICA REGION

Source: World Bank, 2018. DSAT Editor, 2020. Created by author, 2020.

The Middle East and North Africa region (figure 8) contains 21 countries from Morocco in the west to the Islamic Republic of Iran in the east. This region was responsible for generating 129 million tons of waste in 2016. It is important to know that this average was the lowest total of any region, of course, mainly due to its lower population rate. Its average was 0.81 kilogram per capita a day, also very low. According to studies, this region will at least double waste generation by 2050. Here, recycling and composting are being used on a pilot scale. But 53 percent of waste is disposed of in open dumps.

The countries of the Gulf Cooperation Council (GCC) are investing in waste-to-energy projects. But waste generation and management practices can vary due to political issues. Bahrain, Israel, Malta, United Arab Emirates, Kuwait, Saudi Arabia, Qatar, Oman, Libya, and Iraq generate more than 1.0 kilograms of waste per person a day, and countries such as Morocco, Republic of Yemen and Djibouti generate less than 0.6 kilogram per person per day.[38]

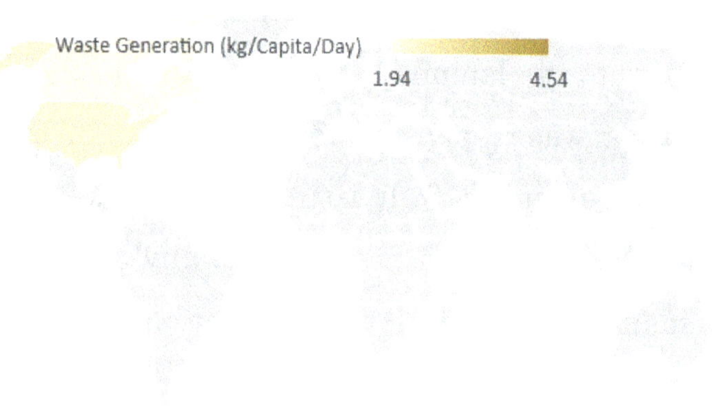

9 NORTH AMERICAN REGION

Source: World Bank, 2018. DSAT Editor, 2020. Created by author, 2020.

[38] WORLD BANK, 2018.

The region called North America (figure 9) is comprised of Bermuda, Canada, and United States, by far the largest country compared to the other two. In 2016, this region generated 289 million tons a year of waste and reached the highest average per capita a day in amount of waste than all regions on our planet. It was 2.21 kilograms a day per capita of waste generated.

Their waste management and disposal practices are more developed when compared to general trends. However, just one-third of waste is recycled in this region although the waste management process works in an environmentally sound way.

The United States generated a waste average of 2.24 per capita a day in that year, a little more that the global average.[39] In 2012, Americans produced about 251 million tons of trash. In fact, 34.5 percent was recycled, therefore, this rate corresponds to about 87 million tons recycled and composted.[40]

---

[39] WORLD BANK, 2018.
[40] Environmental Protection Agency - EPA, 2012.

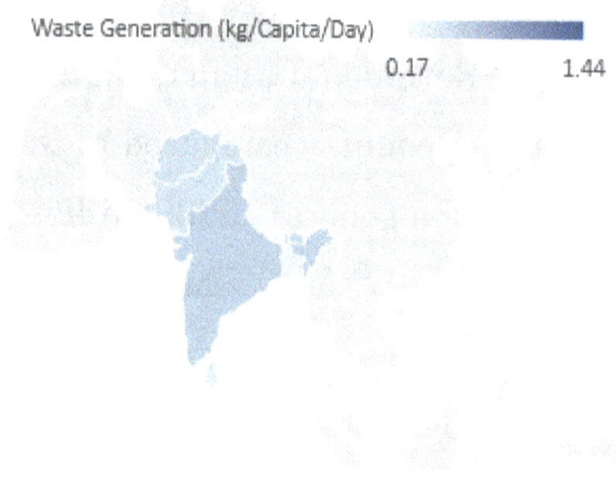

10 SOUTH ASIA REGION

Source: World Bank, 2018. DSAT Editor, 2020. Created by author, 2020.

South Asia (figure 10) includes eight countries with a large population. In 2016, they generated 334 million tons of waste with an average per capita of 0.52 kilogram a day. There is a prediction that this number will reach the double by 2050.

It is a hard fact that to waste management policies are difficult in this region because almost all cities use open dumping. But all 8 governments are working to increase sanitary landfills as well as recycling.

Maldives was the country that generated the most waste in this region. This amount stood at 1.44 kilograms per capita a day.[41]

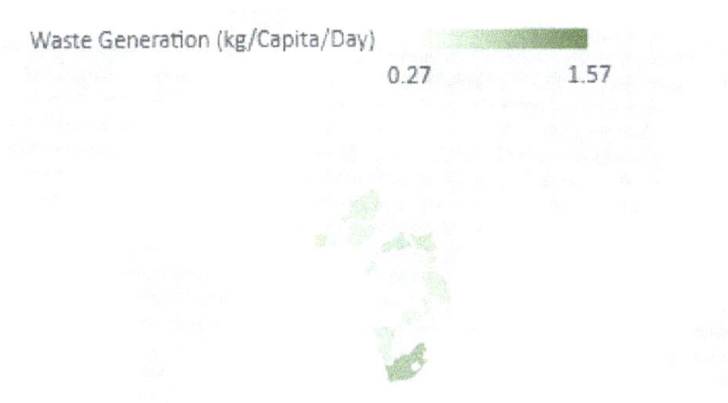

11 SUB-SAHARAN AFRICA REGION

Source: World Bank, 2018. DSAT Editor, 2020. Created by author, 2020.

Finally, the last region to be considered in our observation about World Bank research of 2018, referring to 2016 year, is Sub-Saharan Africa (figure 11). This region has 48 countries and their waste generated that year was 174 million tons, which equates to a rate daily per capita of 0.46 kilogram. Due

---

[41] WORLD BANK, 2018.

to the population growth forecast for this region, by 2050 this amount should triple.

Despite their fast growth, they are working to modernize of waste systems, such as sustainable final disposal places, improvement of collection coverage, dumpsites closed, and development of environmental education. Seychelles and Rwanda generated a waste rate above 1.0 kilogram per capita day in 2016.[42]

About the data presented, it is relevant to further analyze the quality of services and develop metrics to quantify the impact of these services on the environment.

Regarding collection, treatment and, mainly, selective collection and recycling, we can perceive that there is a need for adequacy and expansion of these last services listed aiming at improving processes and reducing the impact on the environment.

---

[42] WORLD BANK, 2018.

In a way, the identification of basic sanitation needs, including the insufficiencies present in low-income countries, has a positive outlook for the acquisition of technical measures aimed at improving the quality of life of the citizens and the environment.

The disposal of urban, agricultural and industrial waste in the air, water and soil gradually makes the available environment increasingly scarce; at some point in order our world will be unable to absorb the increasing amounts of waste.

The sanitary issue, regarding the disposal of waste in less developed places also results in a greater number of dumps in these locations, increasing environmental pollution and consequently affecting the health conditions of this population.[43] Barbosa et al.[44] focuses on the impact, saying that:

> Dumps cause continuous damage to the environment, such as biodiversity impairment; contamination of air, soil and water; emission of gases, risk of fires, increased erosive process, visual pollution and direct risk to public health due to the proliferation of vectors.

---

[43] CALDERAN, 2013.
[44] 2019, apud ALVES, 2020, p. 63.

As regards water pollution, including oceans, seas, rivers, lakes and lagoons, the damage caused is a risk factor not only to human health, but to land, fauna and flora. The impact caused by water pollution can be observed through contamination of rivers, springs, lakes and seas from chemical, physical and biological dumps so harmful to living beings.

Water pollution is also caused by the release of untreated sewage, industrial waste, oil leakage, pesticide contamination carried by rainwater, and other factors that directly or indirectly impact the aquatic environment, water bodies, oceans, groundwater and the terrestrial environment, soil and subsoil.

As a consequence of water pollution, it is documented to aggravate human health, decrease the quality of available drinking water, threaten aquatic life and the survival of organisms, plants and animals both directly and indirectly.

The projected increase in production, consumption, and industrial activity, causes several types of pollution, as we have demonstrated. Pollution is a very serious environmental problem arising from any activity that causes a negative impact on the environment. Through polluting we introduce substances or energy into the environment causing problems for the ecosystem, nature and all living beings.

Lately thankfully! people all over the world are seeking to ensure the preservation of natural resources, including concern about climate change, deforestation, production and excessive consumption, environmental pollution and its impacts.

2.3 Climate change, greenhouse gases and environmental issues

Regarding the impacts suffered by the use of fossil fuels, it is known the harm that planet has been experiencing. Also known are the serious consequences through climate change, the increase in temperature and natural events due to the process of pollution caused by economic development, production and consumption. Global warming from elevated concentration of greenhouse gases, including carbon dioxide (CO2), methane (CH4), nitrous oxide (N2O), contributes to elevating the global temperature of the planet.

Studies point to carbon dioxide (CO2) as one of the main gases responsible for the greenhouse effect. In addition to increasing the planet's temperature, there is acidification of rivers and forests, climate change, melting of the polar ice caps and rising sea levels, acid rain and other proven deleterious effects. Environmental damage causes loss of biodiversity, decreased fauna and flora, and negative effects on human health.[45]

According to figure 12, China appears as the largest emitter of carbon dioxide (CO2). Until 2007, the United States was the largest. The United States emits 17.6 tons of CO per capita and China around 6.2 CO per capita[46], making the us still the worst offender. "We are number one!"

---

[45] SILVA, 2016, apud OLIVEIRA, 2017.
[46] SACHS, 2015.

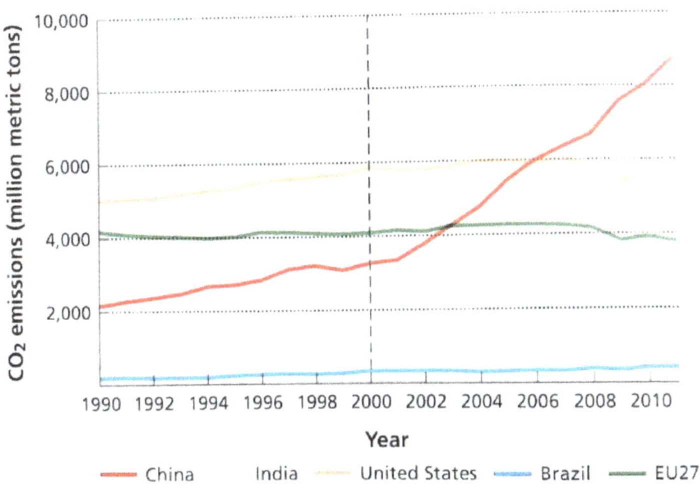

12 CO₂ Emissions

Source: SACHS, 2015. Adapted by author, 2020.

From 1998 to 2018, the fifteen countries that emitted the most carbon dioxide (in megatons) into the atmosphere were China (10,065); United States of America (5,416); India (2,654); Russia (1,711); Germany (759); Iran (720); South Korea (659); Saudi Arabia (621); Indonesia (615); Canada (568); Mexico (477); South Africa (468); Brazil (457); and Turkey (428).[47] In 2007, China overtook United States and became the largest emitter of carbon dioxide on

---

[47] GLOBAL CARBON ATLAS, 2018.

the planet, a position it holds today. In 2018, Brazil was the 14th country that most emitted carbon dioxide (CO2). Mexico was the only country belonging to Latin America in this group until then.[48]

According to The World Bank:

> Carbon dioxide (CO2) makes up the largest share of the greenhouse gases contributing to global warming and climate change. Converting all other greenhouse gases (methane (CH4), nitrous oxide (N2O), hydrofluorocarbons (HFCs), perfluorocarbons (PFCs), Sulphur hexafluoride (SF6)) to carbon dioxide (or CO2) equivalents makes it possible to compare them and to determine their individual and total contributions to global warming.

Figure 13 shows the total emissions (kt[49] of CO2 equivalent) by countries that have generated great pollution by greenhouse gases, as analyzed below according to The World Bank.

In 1971, the United States ranked first of total gas emitters at 5,440,420.6, followed by Russia with 2,244,826.9 and China with 1,914,331.8. In 2012, the last year of this analysis, China was the country

---

[48] BBC BRAZIL NEWS, 2019.
[49] Kiloton: a unit of weight or capacity equal to 1,000 metric tons. (AMERICAN HERITAGE DICTIONARY OF THE ENGLISH LANGUAGE).

that generated most greenhouse gases, the United States ranked second and Russia ranked fifth among these countries.

Indonesia generated 294,004.9 in 1971. Its greatest generation peak was from the years 1982 at 2,488,671.3, and 1997 at 5,040,841.3. In 2012, its generation was reduced to 780,550.8.

In 1971, Japan generated by 960,481.9 and in 2012 its generation increased to 518,377.0, totaling at 1,478,858.9.

Russia, in 1971, was the generator of 2,244,826.9. In 2003 its generation increased to 3,542,027.5, and in 2004 it decreased to 2,413,520.9. In 2012 it generated a total at 2,803,398.5, leaving the second position for fifth in this analysis.

In 1971, Brazil generated 968,552.5, and in 2012 it started to generate at 2,989,418, lagging slightly behind India.

India has seen a significant increase over the years. Its generation of greenhouse gases in 1971 was at 754,018.5, and in 2012 increased to 3,002,894.9.

In 1971 the United States had a total generation at 5,440,420.6 leading the production of greenhouse gases until 2007. In 2012 U.S. generated a total of 6,343,840.5, behind China.

China in 1971 generated 1,914,331.8, and in 2012 increased to 12,454,710.6. From 2004, China ranked second in greenhouse gas generation, behind the United States. From 2008 onwards, China took first place in the generation of greenhouse gases through today.

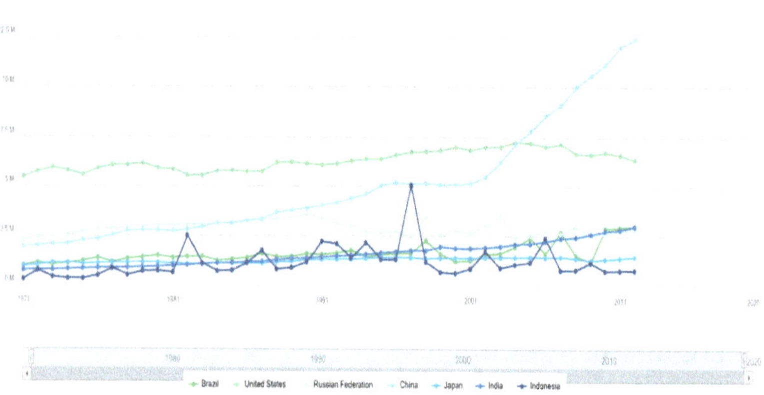

13 Total Greenhouse gas emissions (kt of CO2 equivalent)

Source: WORLD DEVELOPMENT INDICATORS. Created by author, 01/16/2021.

The production of greenhouse gases disturbs the earth's radiative balance. It has led to an increase of surface temperature and harmful effects on climate, sea level rise and agriculture.[50]

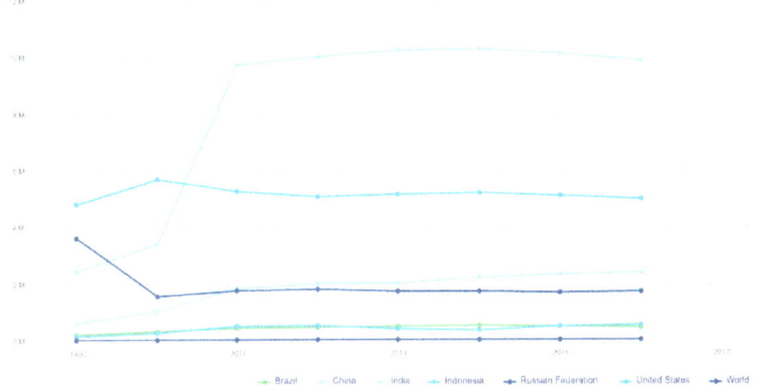

14 CO2 Emissions (kt)

Source: WORLD DEVELOPMENT INDICATORS. Created by author, 01/16/2021.

Further illuminating the increase of pollution levels over the years, figure 14 presents more recent data. In 2016, China remained at the top of CO2 emissions with a generation of 9,893,038.0, ahead of the United States with 5,006,302.1. Following the same ranking worldwide place, India

---

[50] WORLD BANK.

has occupied the third place from before with 2,407.671.5 in CO2 emissions (kt).

Air pollution is known to increase due to industrial development impacting urban air quality. High concentrations of polluting gases in cities are caused by industrial activities and processes, the release and combustion of gases from motor vehicles, dispersion of suspended solid particles and other forms of discharge of toxic gases into the environment. Therefore, urban pollution directly affects the air quality and life of people, animals and nature in general, including rivers, lakes and seas.

The emission of vehicular gases contributes directly to air pollution, mainly observed in large urban centers, with pollution rates similar to those of industries.[51] According to Environmental Protect Agency (EPA, 2008) the average vehicle using gasoline generates 4 tons of carbon dioxide (CO2) per year. Much of the air pollution in large regions of large cities comes from motor vehicles. Other problematic sources

---

[51] BRIENZA et all, 2015.

are industries and fires, aggravated by worsening weather conditions.

2.4 Deforestation and environmental issues

Deforestation is another important factor to be considered as a major global impact. It is well-known that the balance of the planet also comes from forests and biodiversity included in this ecosystem.

Economic development and land use for certain purposes stimulate deforestation. Some activities on this end may be agriculture, livestock, industrial and commercial use of timber, mining, the opening of roads for the transport of resources and even housing.

It is worth mentioning the impact caused by deforestation, which includes the interruption of the food chain, the extinction of species and the elimination of their habitat, among other impacts with the burning and carbon emissions resulting from this activity.

According to the United Nations Environment Programme (UNEP[52], 2020), statistical data indicate globally that the livelihoods of approximately 1.6 billion people depend directly on forests. It is well-known that forests concentrate a large majority of all species. As a result of forest degradation around 8 percent of terrestrial animals are extinct and 22 percent at risk of extinction around the globe today.

Figure 15 below shows the impact on biodiversity through deforestation. Vegetation loss can be observed by comparing the previous vegetation cover (former) and current vegetation cover. One can note the impact caused to forests, and especially the Amazon Forest in the area indicated in the figure as tropical, which encompasses part of the equatorial tropical forest, and where the Amazon basin is located.

Humanity has been working to fight deforestation for thousands of years, but even today

---

[52] Leading global environmental authority within the United Nations system that defines the global environmental agenda, promotes the coherent implementation of the environmental dimension of sustainable development and acts as an advocate authority for the global environment.

due to human pressures we continue to lose forest areas.[53]

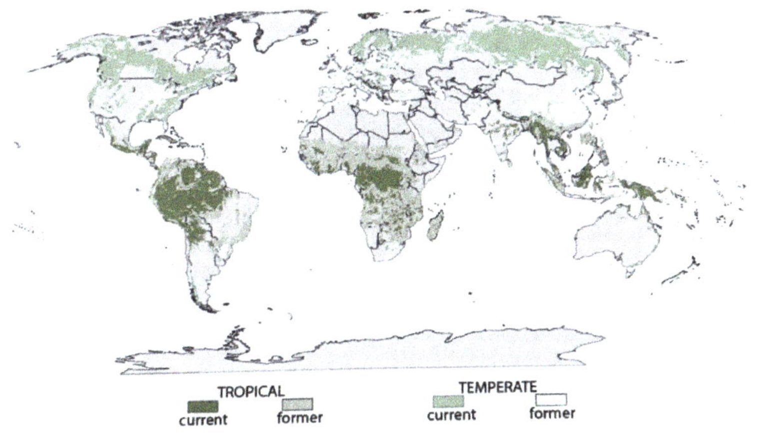

15 Global Distribution of Original and Remaining Forests

Source: SACHS, 2015. Adapted by author, 2020.

The Amazon has become the focus a great number of international environmentalists, scientists, teachers, critics, activists and others in support of the struggle for environmental balance, far beyond governments.

As discussed earlier, the environmental crisis caused by the current development model

---

[53] SACHS, 2015.

through the use of fossil fuels, as the main source of energy, has been a major global concern.

In view of the above, we clearly see the need to reduce these pressures on environmental resources. For that, it is necessary to reduce socioeconomic inequalities and thus to promote a search for social justice and environmental protection.

The unsustainable consumption of natural resources results in the degradation of the physical, biological and social systems and is related to the increased risk of public health problems.[54]

The relationship between human society and the environment has been affirmed as one of the main concerns, both in the field of public policies and in the field of knowledge production.[55]

In the view of the social sciences, especially anthropology, it is necessary to understand the complexities of the environment, addressing it from a global and interdisciplinary point of view. This statement is based on the fact that anthropology was

---

[54] FORATTINI, 1992.
[55] FOLADORI, 2004.

born by asking about the anthropic transformation that different societies produced in their environments, the continuity and difference of the human species in relation to other living beings, and also the place of consciousness in social evolution.

The environmental crisis is a result of technical development. Some contemporary environmental movements and many ecodevelopmental authors focus their criticism of the environmental crisis on technological and industrial development. They start from the supposed, often unexplained, autonomous evolution of technique and technology, a linear evolution from simple instruments to complex machines, parallel to the alienation of men with respect to working instruments and the environment.[56]

Environmental problems are objective and must be scientifically analyzed. Before the mid-1980s, environmental problems were national, regional or local; they were related to river contamination,

---

[56] FOLADORI, 2004.

deforestation, urban environmental pollution, the depletion of animal and plant species, the effects of chemicals on health, etc. From the mid-1980s, climate change became the common denominator of all environmental problems, and global warming the main defendant.

Everything is related to climate, and the reduction of global warming has become the goal of international environmental policy. Uncritically, many environmental organizations and groups have agreed to regard global warming as responsible for the environmental crisis.

Climate change reflects the relationship of every aspect within the whole. It affects biodiversity, has an impact on the health of forests and suffers the effects of it, affects human productive activity and, is connected to many infectious diseases, etc.

Recently, a concern with the scarcity of natural resources has given a new dimension to the environmental conversation. This perspective gained worldwide prominence with the proclamation by the United Nations of the Year of the Environment, 1970, and with the convening, also by the United Nations, of world conferences on the environment.

In the early 1970s there were two polarizing positions on the environment. One, expressed in "The limits of growth", of the Club of Rome, proposed an immediate halt to economic and population growth. Another, expressed in the declaration of the Stockholm Conference held in 1972, proposed the correction of environmental damage caused by economic development and the stabilization, in the medium term, of world population.[57]

Fifty years later, by 2020 the planet experienced the climate crisis through the COVID-19 pandemic, both collapsing modern society and causing numerous ripple effects around the world.

On one hand, the environmental crisis and the growing industrialization are observed as interdependent complexes related to the increase in average global temperature, caused by urban growth, the force of the economy and inequitable power

---

[57] VIOLA, 1991.

distribution. Individuals fell ineffective in the face of a serious problem; governments have relaxed more effective controls on environmental pollution in favor of economic growth and they have influenced flows of large oil, coal and timber industries, among others.

On the other hand, the entire world population is experiencing an alarming numbers of deaths and long-term disabilities from COVID-19 disease throughout the world. This has caused fear, death, unemployment, hunger, social withdrawal and still unknown impacts on the world economy.

Although vaccination rates have been increasing, many countries are still suffering the consequences of lack of access to them. In addition there are denialists and disbelievers of the disease who contribute to fake news. Despite all this pain, the world needs to evolve and contribute to science in decisions about the crises that the planet is going through, both environmental and public health.

Thus, as we delimit and contextualize the concept of environment as relations between humans and nature for the preservation of natural resources in processes of sustainable development, we can understand the urban environment as a relationship between people with constructed space and nature in concentrations of population and human activities.

Constructed space is the result of the profound transformation of the environment to adapt it to the needs of large groups of people, suitable habitat for human activities. We need, however, to define the nature of the constitutive relations of the environment from the perspective of sustainability in a definitive and growing way.

3

SUSTAINABILITY

3.1 Sustainable development

To discuss sustainable development it is necessary to understand exactly what it means. According to Brundtland (1987), sustainable development is development that meets the needs of the present without compromising the ability of future generations to meet their own needs.[58]

Historically, sustainability emerged as a point of support for the scarcity of natural resources, mainly due to population growth in the 18th and 19th centuries. In the second half of twentieth century, when industrial development was positioned in the market to cause significant environmental impacts, the concern with this subject gained space.

---

[58] SACHS, 2015.

Initially, sustainable development sought to integrate environmental issues with economic policy.[59] The impact of pollution in the 60s and 70s served as a basis for international environmental policy. In the last decades, sustainability has reached every conversation due to environmental concerns and policies, with advances and setbacks throughout the history of sustainable development.

The concept of sustainable development imposes itself as a new semantic element of international language and a focus element of the work of experts from international organizations.[60]

Sustainable development aims to guarantee economic growth essential to meet the current demands as well as the needs of future generations through the rational use of natural resources for the benefit of social well-being. It must be understood that broadly includes three important components: economic development, social inclusion, and environmental sustainability.

---

[59] DRESNER, 2002.
[60] HATEM, apud GODARD, 1997.

The complexity of sustainability increases the need and importance of actions from all sectors of environmental management to seek integrated and sustainable solutions.[61]

This complexity translates into planning systems that do not accompany the dynamicism of cities, the complexity of the environment and their role in the development process, discontinuity of policies, plans programs and projects. Systems will reduce or cancel the positive impacts of investments in infrastructure and operationalization of human settlements. Finally, the low participation of society in this management process.[62]

Environmental changes imposed by the significant consumption and production patterns of societies considerably alter natural environments, polluting the physical environment, consuming natural resources without adequate metrics, increasing the risk of exposure to diseases, and generally acting negatively on the quality of life.[63]

---

[61] MALHEIROS, 2005.
[62] MALHEIROS, 2005.
[63] MIRANDA, 1994.

The main concern of sustainable development in practical terms may be related to our crowded planet and its ability to support self-recovery naturally. As already posed, this means a surcharge on the use of natural resources. When humanity reaches the limit overtaking the level of natural resources that can be renewed, the planet enters a kind of negative credit. This resource deficit started in the 70s. We well-know these subcategories now: deforestation, soil erosion, drought, food shortages, wildfire, and other drastic consequences caused by climate change that will be discussed later.

Environmental threats are arising on several fronts and changing Earth's climate.[64] This alarming situation is due to the use of greenhouse gases caused by industrialization, accelerated population growth and disordered consumption of resources.

The planet we want from the perspective of sustainable development must be the focus of

---

[64] SACHS, 2015.

humanity's actions. To achieve the common good, we must act with attitudes based on the preservation and conservation of the planet, including economic and social relations. After all, the term "sustainable development" was created years ago and continues to be so current provoking reflections about sustainable politics in all countries that are seeking a better planet.

3.2 Sustainability policies

As a new way of developing economically, socially and environmentally, countries need to be attentive to adequacies of natural resource management. This implies seeking new ways to promote human well-being, economic development and preservation of natural capital, without being used or degraded, as if it were worth nothing. The challenge of fighting poverty must be tackled, while keeping environmental costs front and center.[65]

---

[65] CAVALCANTI, 2002.

Clearly, a development strategy should not be based simply on the predatory form of the use of nature without considering a significant compensation for the loss of natural capital in which it occurs.[66] Thinking through the perspective of natural capital, as a source of income and wealth, all plans must be rational and sustainable. The use of natural resources must be in accordance with all environmental sustainability policies provided by local, regional, federal and international laws.

Regarding the formulation of public policies for sustainability, we emphasize that the real environmental problem consists precisely in raising the value of nature, using its stocks healthily, without overloading the functions of supply, source (resources) and absorption or fossa (waste) of the ecosystem.[67] Moreover, using natural resources respecting the planet's ability to support its natural self-recovery depends on curbing some economic development. That issue has recently been discussed in local and

---

[66] CAVALCANTI, 2002.
[67] CAVALCANTI, 2002.

international forums in the search for the balance of the planet. We must stop treating (water, air and) soil like dirt!

Sustainable development policies since its emergence always mean, irrefutably, some form of environmental degradation. Now we have to use nature in a more lasting, sober and healthy way than has been the practice to date.[68]

It is undeniable that in the discourse on development, it is always the goal to increase per capita income indefinitely. This would be laudable if this model involved a true and permanent reduction of poverty.[69]

The problem becomes it is empirically verified that increasing amounts of natural capital are needed to produce each marginal unit of poverty reduction, as is proven by study after study on scarcity.[70]

To choose sustainability means to adopt an orientation to conserve more natural capital for future

---

[68] ROEGEN, 1974, apud CAVALCANTI, 2002.
[69] DALY, 199, apud CAVALCANTI, 2002.
[70] CLEVELAND, 2002.

generations. This implies acceptance of a philosophy of finitude and self-restraint, which is not easy to reconcile with globalized attitudes of consumption.[71]

An important principle of policy formulation for sustainability is to have a consistent information system to measure the economic performance of a country or region. In a sustainable society, progress must be quantified with metrics on quality of life (health, longevity, maturity, psychological, education, a clean environment, community spirit, leisure enjoyed intelligently, and so on), and not simply on material consumption.[72]

Government policy for sustainability means orientating public actions on the fundamental ecological limitation of resources, without which no human activity can be carried out. The strategic problem is to find the sustainable metabolic flow that can elevate social well-being without causing damage to environmental functions and services. In other words, the level of the social product must be

---

[71] CAVALCANTI, 2002.
[72] VIOLA, 1996.

guaranteed in the same way as the quality of the natural environment and the quality of life.[73]

Finally, the punctual obstacle all governments confront in complying with laws and executing plans and programs aimed at sustainable development is called "the inertia" of environmental policies.

---

[73] ROEGEN, 1971.

4

CARTESIAN PARADIGM AND ITS INFLUENCE

The existence of the human species is immersed in a crisis that can be seen in economic, religious, political and social aspects, and endangers the future of human kind.

The self-destructive madness of the human being seems to be limitless. Anything that pleases the ego is superior to the considerations of the consequences of our actions. The struggle for supremacy in fierce competitions to make quick profits no matter the cost or the possibility of breaking ethical standards, is put ahead of environmental considerations. Man becomes increasingly self-centered and renounces nature.[74]

In a few centuries man has destroyed and endangered countless animal and plant species.

---

[74] PECCEI and IKEDA, 1984.

Cities, metropolises, megalopolises were born, justified by development. Man has violated and destroyed nature in all parts of the world through technological advancement and demographic explosion.[75]

Science and technology are dynamic today. This dynamism is due to the Industrial Revolution and Cartesian principles. The mastery of nature by man has relied on the costs of slavery to an artificial and inadequate environment. Man dares to enjoy his artificial creation in order to seek happiness.

A human life is necessarily interdependent. It is our collective responsibility to share in the construction of universal harmony. However, this way of thinking about the relationship with the environment is fairly recent.[76]

Civilization, especially in the West, in the when formulated by Bacon and Descartes leads to unknown paths. Technology and science become imperative, accentuating ambition and perpetual

---

[75] TOYNBEE, 1974.
[76] TOYNBEE, 1974.

growth. This behavioral and existential formula generates an identity crisis that creates a broken link in the man-nature relationship. Man loses the notion of existence and moves toward a mechanical notion of being. Detachment from nature and unmeasured affection for reason and technique, driven by Cartesian thought, makes it predators and explorers.[77]

Nature opposes this binary new and its defeat turns into the defeat of humanity when in pursuit of absolute power. In turn paradigms of macho dominance are created. These ultimately weaken the man-nature relationship, and permeate all spheres of Western civilization.[78] Mother Earth has been offering all her riches to the delight of science and technology, coinciding with these macho views toward women.[79]

Newtonian science and its mechanistic ideas of nature; a machine that can be explored and manipulated. This perpetuates the path of female exploitation. The link between women and nature

---

[77] TOYNBEE, 1974.
[78] LA TORRE, 1993.
[79] CAPRA, 1982.

helps grow feminist and ecological movements worldwide.[80] This model is 100% contrary to the macho and anthropocentric Cartesian one.

The roots of the environmental and human crisis are presented by Francis Bacon; by creating conditions for him to improve his own living condition, he gives himself dominion over the environment.[81]

With Descartes comes the synthesis of the anthropocentric view. For Western rationality this begins to guide human attitudes. This overvaluation of physical being makes man disconnect from the collective and intuitive senses, proper to the feminist conception. This is turn cause a dislearning of the relationship with the environment. Nature is then considered an exploitable object.[82]

The author also states above that the excessive emphasis on rationality and the scientific-technological win-lose method leads man to antiecological attitudes. And he suggests that, in order to have a balance of ecological consciousness, it

---

[80] CAPRA, 1982.
[81] LA TORRE, 1993.
[82] CAPRA, 1982.

is necessary to combine linear knowledge of technology and science with the intuitive nonlinear knowledge of indigenous peoples. For these, by conceiving the relationship between man-nature in a holistic and interdependent way, it can lead to a new form of relationship with the environment.

Cultural and social evolution causes the loss of biological and ecological references. Intellectual development, scientific and technological knowledge depart from wisdom and ethics. These freeze, as if they were finished in themselves. Rational and intellectual progress is an alarming place in the face of possible self-extermination.[83]

In his concept of progress, man distakes the powers of the natural elements, as if they were in no way threatened. With a technical-scientific theory of domination, he believes that everything can be controlled rationally. This uncommitted relationship with nature, creates an anthropocentric culture, and

---

[83] CAPRA, 1982.

nature, unknown and feared, languishes, without the need for care.[84]

Anthropocentric ethics characterize all non-human objects as having no relevance; traditional ethics is anthropocentric and man is seen in his condition as an essence, to be present in the world and not an object of transformation; and all the relationships faced by the human being in the man-man relationship are worthy of ethical and moral judgment, but this judgment does not apply to the man-nature relationship.[85]

However, more modern thought introduces new elements. All these changes impose on ethics the dimension of responsibility. The first and greatest change is the vulnerability of nature. This vulnerability makes man responsible for the biosphere.[86]

Ethics should concern the rights and duties of each of us in our interrelationship with all creation.

---

[84] CAPRA, 1982.
[85] JONAS, 1995.
[86] JONAS, 1995.

Your role is to become eco-ethic; focus on a balanced relationship between all ecosystems and the self.

If man places the biosphere under his tutelage, it becomes necessary to think of an extrahuman ethic, which benefits not only human beings, but also nature itself. We therefore expand the recognition of ends in themselves beyond the human sphere and incorporate into it the concept of our planet's care.[87]

In times past, technology was used simply to meet human needs. Nowadays, technology has an infinite impulse aiming at an increasing dominance over things and over men themselves. Man is increasingly the product of what he produces. For him, there is no self-contradiction in the idea that humanity ceases to exist. An imperative that perfectly fits the new style of human actions is that the effects of these actions are compatible with the permanence of human life on Earth.[88]

---

[87] JONAS, 1995.
[88] JONAS, 1995.

This new imperative must be addressed by public policy. It calls for agreement of the lasting effects of the individual for the continuity of human activity in the future. Rulers have responsibility for today's humanity and the well-being of future generations. Once creating ethical conditions of coexistence in society, good leaders will be contributing to citizens seeking high values outcomes, predisposing themselves to sacrifices for the common good.[89]

It is important to emphasize that the object is not questioning the validity of ethics for its own purposes, but its insufficiency in the face of new human actions, namely, technology and science. That is why an ethics of foresight and responsibility is necessary as new circumstances arise. And, it is up to the reformulated and holistic ethics to constantly revise and update useful standards of conduct.[90]

Also according to the author above, human knowledge has given man, scientific forces that need

---

[89] JONAS, 1995.
[90] JONAS, 1995.

to be regulated by norms. We must establish an ethics that can curb extreme capacities, because ethics exists to contain man's actions and regulate his power. Therefore, the new capacities of action require new ethical rules, or perhaps an entirely new ethics.

By thinking about responsibility for future generations, one cannot conceive it in the traditional way of rights and duties. It is a principle based of equilibrium of relations between human beings with each other and with nature.[91] This author explains a scenario in which the rights and duties of an anthropocentric ethics do not prevail, but rather the spontaneous desire to contribute to the existence of future generations does. It is a solid, fraternal responsibility of natural and creational and non-personal merits. Thus, it is necessary to resuscitate altruism, and understand it as part of our natural identity and interdependence.

The West does not allow itself to fully embrace the idea of its non-responsibility to nature,

---

[91] JONAS, 1995.

through only conservation and preservation. It is this sleeping seed of responsibility that must be reawakened. It is our shared moral responsibility, which goes beyond the norms of rights and duties. It is our collective responsibility that only can occur if there is a universal identity, providing a balanced and harmonious ecosystem.[92]

Then, two forms of ethics are distinguished: the simple ethics of duty and responsibility, and a more comprehensive ethics of aspiration. In the ethics of duty are respect for nature, preservation, responsibility for future generations. In the ethics of aspiration, feelings of beauty, harmony, mystical and fraternal union with nature are considered, the relationship between care with creation.[93]

Care means zeal, solicitude, diligence, attention, good treatment. Thus, nature is not seen as an object of exploration and domination. There's an on-going relationship of coexistence. It ceases to be

---

[92] PASSMORE, apud LA TORRE, 1993.
[93] LA TORRE, 1993.

subject-object and passes the condition of subject-subject.[94]

When we consider care, we cease seeing otherness. It is the absolute ethical untouchability. One-ness in the human-nature relationship happens when man strips himself of the totalizing, centralizing and exploitative position, and seeks a relationship of respect, holistic and mutual responsibility.[95]

One of the great challenges for the way-of-being-care is the way-of-being-work, are two complementary dimensions, but understood and experienced in opposite ways. Work is a way to accumulate capital, the conquest of the other, of the world, of nature. On the other hand, the way-of-care springs from feeling, understanding, empathy, care. And this feminine way of being makes nature Mother and Sister.[96]

This new position does not mean a total abandonment of work or reason, but a form of

[94] BOFF, 2000.
[95] DE SOUZA, 1996.
[96] BOFF, 2000.

balanced and harmonious integration with the universe.

An ethics of responsibility is a form of solidarity with all beings.[97] He postulates a mystique which leads human beings to a fraternal and solid reconnection. The author makes clear the existence of two ecologies, the outer ecology, represented by ecosystems, and the interior, represented by solidarity.

The ethics of responsibility associated with the ethics of compassion generates a new understanding of nature. Constructive bonds are formed, in which care is the driving force of action. Reason, in turn, becomes an instrument of ethical distinction, assisting in choices. This new ethic could generate a new way of being in the world and establish a new alliance between reason, compassion and responsibility.[98]

---

[97] AUER, apud BOFF, 1999.
[98] KÖSEL, apud BOFF, 1999.

5

GOVERNMENT, SUMMITS, AGREEMENTS AND SCIENTIFIC RESEARCH: THE B SIDE OF HISTORY

The multiplication of accidents and environmental problems and the action of the ecological movement, from the 1970s on, are a critical force affecting the models of industrial development, both capitalist and socialist. This awakened a new awareness to the environmental dimension of reality.

This new awareness of organized civil society led to the incorporation of environmental policy into national government programs, the party-political system and the agenda of international organizations. The International Conference for the Human Environment, promoted by the United Nations in 1972 in Stockholm, Sweden, is the historical-political framework of a series of national and international initiatives and events dealing with environmental issues.[99]

That same year, the controversial "The Limits of Growth" was published by the Massachusetts Institute of Technology - MIT, on request from the Club of Rome. This document assessed the conditions of planetary environmental degradation and set out predictions for the future. The published results were starkly pessimistic: either there is a change in economic growth patterns, or ecological collapse will occur within the next hundred years. The debate polarized groups who proposed zero growth and those who did not trust such predictions and believed in the potential of science and technology as a way out of the impasse.[100]

At the Stockholm Conference, UNEP - The United Nations Environment Programme - was created. This Conference highlighted the concern with the vulnerability of natural ecosystems, emphasizing the technical aspects of contamination caused by

---

[99] GUIMARÃES, 1991.
[100] GUIMARÃES, 1991.

accelerated industrialization, demographic exploitation and the expansion of urban growth.[101]

The agenda of this Conference presented themes and objectives of interest to industrialized countries. Brazil, on this occasion, was one of the countries to lead the third-world resistance, on the grounds that economic development superseded environmental control; if pollution results inevitably, it will come. The fact that developed countries proposed control over economic growth after reaching high rates of growth and degradation of their own resources was anathema.

Despite the conflicts, the Conference has since resonated worldwide, triggering other international conferences, the creation of several international agencies concerned with the theme, environmental agencies and ministries in most countries, thousands of non-governmental organizations and the organization of green parties in several countries.

---

[101] GUIMARÃES, 1991.

In 1973, the concept of ecodevelopment was used to characterize an alternative conception of development. The basic principles of this new development perspective were formulated by Sachs and could be synthesized as the satisfaction of the basic needs of the population, solidarity, participation, preservation of natural resources and environment, development of a social system that guarantees employment, social security and respect for other cultures and education programs.[102]

This concept emphasizes opposition to imitative growth models, the import of inadequate technologies. It promotes the autonomy of the populations involved, in order to overcome the cultural dependence on external references.[103]

In 1974, another landmark document in the debate on development and the environment, is the Cocoyoc Declaration. This was the result of the meeting of the United Nations Environment Programme - UNEP and the United Nations

---

[102] BRÚSEKE, 1996.
[103] VIEIRA, 1995.

Conference on Trade and Development - UNCTAD. At this meeting, progress was made on the model suggested by Sachs, leading to discuss on of: the connection between population explosion, poverty, environmental degradation, and the responsibility of developed countries, due to their high level of consumption, waste and pollution.[104]

The Que Faire Report, presented in 1975 by the Dag - Hammarskjöld Foundation with the participation of researchers and politicians from 48 countries and contributions from UNEP and 13 other UN organizations, reinforces Cocoyok's arguments. Criticism of the abuse of power of developed countries, the excessive interference of these countries in the fates of third world countries and the environmental consequences were presented. For Vieira (1995) hopes were placed on development strategies based on the self-confidence and autonomy of poorer countries. This radical view suffered serious resistance from conservative governments, scientists and politicians.

---

[104] VIEIRA, 1995.

In 1983, the World Commission on Environment and Development (UNCED) was established by the United Nations General Assembly. This Commission, under the Presidency of Gro Harlem Brundtland (Norwegian Prime Minister), aimed to re-examine the main problems of global environment and development, and formulate realistic proposals as solutions. The report "Our Common Future" was presented in 1987, and was based on the assumption of the need to reconcile economic growth and environmental conservation, disclosing the concept of sustainability.[105]

In this context, sustainable development was used for the first time in 1987 in the Brundtland report by World Commision on Environmental Development. This concept, already defined earlier here, seeks a commitment to development from the understanding of the needs of future generations to be met, as well as the needs of current ones. The Brundtland report was innovative in that it refused to

---

[105] GUIMARÃES, 1991.

deal exclusively with environmental problems, opting for a balance between development styles and multidimensional articulation involving economic, political, ethical, social, cultural and ecological concerns.

This report presented a philosophy of development combining economic efficiency with ecological prudence and social justice. It also emphasized that the problems of the environment and sustainable development are directly related to the problems of poverty, satisfaction of basic needs, food, health and housing and an energy matrix that privileges renewable sources in the process of technological innovation.[106]

In 1992, twenty years after Stockholm, the United Nations Conference on the Environment - UNCED, known as the Earth Summit or Eco 92, 178 countries met and parallel to this event, the Global Forum with about 4,000 NGOs took place in Rio de Janeiro. As Barbieri (1997) reported, an Earth

---

[106] GUIMARÃES, 1991.

Charter was foreseen, with declarations of the fundamental principles of sustainable development. 27 principles that expanded those of Stockholm was approved, and was called the Rio de Janeiro Declaration on the Environment and Development, another landmark document.

According to Sachs (2015), one of the key principles of the Rio Declaration was that "development today must not threaten the needs of present and future generations".

Rio - 92, the United Nations Summit on Environment and Development, strengthened the search for sustainability, implemented Agenda 21[107] and proposed discussions on the Climate Convention. This conference adopted two important multilateral environmental agreements, the United Nations Convention on Climate Change and the Convention on Biological Diversity (CBD).[108] Also according to this

---

[107] Agenda 21 is a comprehensive plan of action to be taken globally, nationally and locally by organizations of the United Nations System, Governments, and Major Groups in every area in which human impacts on the environment. Agenda 21, the Rio Declaration on Environment and Development, and the Statement of principles for the Sustainable Management of Forests were adopted by more than 178 Governments at the United Nations Conference on Environment and Development (UNCED) held in Rio de Janeiro, Brazil, 3 to 14 June 1992. (https://sustainabledevelopment.un.org/outcomedocuments/agenda21).

[108] SACHS, 2015.

author, this event was presented as a basis for another discussion, which would be held two years later, called the United Nations Convention to Combat Desertification.

In 1992, at the Rio de Janeiro Earth Summit, sustainable development was discussed comprehensively. At that time, sustainable development was the focus of international concern in establishing policies capable of reconciling economic and social development with the balanced use of natural resources. During that meeting, it was decided that developing countries should receive technological and financial support of developed countries aiming to reduce consumption patterns, priority fossil fuels, the oil and coal industries.

Ten years after Rio-92, Johannesburg, in 2002, the stage was set for the UN World Summit on Sustainable Development. In that summit they planned the implementation of the three components of sustainable development: economic development, social development and environmental protection.[109]

Despite global criticism for the lack of tangible results in environmental protection since Rio-92, Johannesburg Summit was considered a landmark in international diplomacy regarding the environment. It resulted in important decisions about future meetings. The most relevant ones would deal with biodiversity and global climate change.

The Future We Want was the theme of the Rio + 20 Conference in 2012 in Rio de Janeiro - Brazil. The United Nations Conference on Sustainable Development, informally called the Rio + 20 Summit, was a global meeting where social inequality was discussed with a view to raising the standard of living of humanity.

At the 2012 meeting, world leaders concluded in a final document called "The Future We Want" saying absolutely that we should not give up:

> We are committed to reinvigorating the global partnership for sustainable development that

---

[109] SACHS, 2015.

we launched in Rio in 1992. We recognize the need to give new impetus to our cooperative pursuit of sustainable development and are committed to working together with key groups and other stakeholders to address implementation gaps.[110]

By the way, The Future We Want reaffirmed the need to achieve sustainable development by:

"Promoting sustained, inclusive and equitable economic growth, creating greater opportunities for all, reducing inequalities, raising basic standards of living; fostering equitable social development and inclusion; and promoting integrated and sustainable management of natural resources and ecosystems that supports inter alia economic, social and human development while facilitating ecosystem conservation, regeneration and restoration and resilience in the face of new and emerging challenges."[111]

In addition, globalization, environmental impacts and the poor distribution of income were raised. This discussion highlighted the need to eradicate poverty and change consumption patterns to achieve sustainable development.

---

[110] UN General Assembly, 2012 apud, SACHS, 2015.
[111] UN General Assembly, 2012 apud, SACHS, 2015.

In 2015, New York hosted the Sustainable Development Summit, where the new Sustainable Development Goals (SDGs) and a universal sustainability agenda for 2030 were defined. Ending poverty, protecting the environment and combating climate change are the specific objectives of this meeting. Yes, these narratives had already been on the agenda of discussion in 2000 with the adoption of the Millennium Development Goals (MDGs). They also included the fight against extreme poverty by 2015, as a commitment made by the 191 leaders of the United Nations Member States.

Sustainable development is a normative, which means that it recommends a set of objectives to which the world must aspire. The nations of the world have adopted the SDGs precisely to assist in the future course of the economic and social development of the planet.

Finally, in 2019, another major Conference in New York discussed climate change. At this meeting, according to the UN, the main goal was to get the world to achieve carbon neutrality by 2050.

Achieving carbon neutrality means that countries should no longer emit polluting gases than nature is not able to absorb. However, there is alarming chaos in the world that amplifies the climate crisis. The emission of toxic gases being released on a large planetary scale being reduced by 2050 seems ambitious at best.

How do we change the way the world currently grows? This should be the number one issue. It is known that growth contributes to the systemic advance of pollution. In this context, what pollutes the most is the generation of energy, transportation, extraction and burning of fossil fuels. Therefore, there is an urgent need to address the development of a new kind of economy/ economic growth model.

The Summits warned the world about climate change and industrialization as escalating concerns for the future of the planet.

There is no more time for climate-unfavorable rhetoric, agreements not fulfilled by governments, scientific facts discarded, or under investment. The climate is emerging a problem for everyone. Moreover, to resolve this issue, joint actions are needed now to collaborate with the future of the Earth.

Current ecological micro-collapses and social crise are affecting the development of societies. According to Sachs (2015), there has never been a global problem in the economy like the climate issue. This is the most difficult public policy problem facing humanity.[112] This is an urgent cause, a great threat to the planet.

Sustainable Development has reached national and international spaces in a broad way among governance mistakes and successes. The Sustainable Development Goals Report 2020 on the 2030 Agenda for Sustainable, analyzes the progress made from the perspective of the current COVID19 crisis and its impacts on society.

The Brundtland Report has occupied a prominent position in the recent debate on the environment in its relationship with economic and

---

[112] SACHS, 2015.

social development. By incorporating multidimensional perspective, it overcomes unilateral and reductionist approaches to the problem. It also innovates by valuing the problems of north-south relations, and especially the specificities of the poorer countries. It addresses the adverse implications of external debt, recognizing north-south inequality and the greater relative responsibility of northern countries in the construction of sustainable development.[113]

In the field of scientific knowledge, the proposal suggests an approximation between the natural and social sciences.

For those who criticize the concept, the emphasis is on ambiguities and contradictions. According to Lima (1993), the main attacks branch around the reconciliation between economic growth and environmental preservation, in the context of a capitalist market economy.

---

[113] BRÚSEKE, 1995.

The critics also consider sustainable development only a "old wine in a new bottle". They point out the lack of consensus on what is and how sustainable development should be achieved, and what must be done to achieve economic efficiency, ecological prudence and social justice in the context of an unequal, unfair and degraded world.

Development and sustainable are contradictory terminology.[114] To him, development is understood as growth; it means a physical or material production increase, while sustainable means continuity of without growth.

The association of the conceptions of sustainability and development contains an inherent antagonism.[115] He states that sustainability is a concept of ecology, meaning a tendency to stability, dynamic balance and interdependence between ecosystems, whereas development concerns the growth of production media, the accumulation and expansion of productive forces.

---

[114] TENORIO, 2006.
[115] HERCULANO, 1992.

Also according to the author above, economic dimension of development is overvalued and ethical, cultural, social and political aspects, undervalued. For environmentalists, this concept has been distorted and reinterpreted as a strategy for expanding the market and profit, when, in fact, it means essential changes in the production/consumption structure, a new behavioral ethics and the rescue of collective social interests.

Stahel (1995) questions sustainability in the context of capitalism. He argues that the concept has been separated from its capitalist meaning or shows its possibility of realization or will fall into emptiness, serving as a legitimizing ideology of capitalist unsustainability. He develops an analysis from the notion of entropy, looking for the tuningbetween economic and biophysical rhythms. It comms on the two perspectives, observing that biophysical time is guided by stability, continuous recycling and low levels of entropy, while the economic time introduced by capitalism is marked by constant expansion, market competition, constant innovations and instability.

Acceleration of time, characteristic of capitalist logic, breaks with circular time and biospherical stability, accelerating the processes of entropic degradation. And, it is in this time misstep that is the source of the environmental crisis. The capitalist development model, seen from the logic of entropy, proves unsustainable and thus the discourse of sustainability in the context of a market economy is an illusion.[116]

Another controversial point concerns relations between the northern and southern countries. Studies in industrialized countries advocate sustainable development for southern countries, while it is the northern countries that most need it.[117] This author, investigating projections of population growth and fuel consumption in the two blocks, concludes that even with the largest population growth in southern

---

[116] STAHEL, 1995.
[117] ALMINO, 1993.

countries, northern countries remain responsible for most of the global environmental damage.

The Brundtland report avoids referring to sustainable development as an expression of liberal capitalism. It points to poverty as the main cause of environmental degradation. Yet the Brundtland report does not say that both are products of a growth model that prioritizes capital expansion and not human needs.[118] This author identifies the trend of the report as ecocapitalism. She believes technological advances are now capable of generating a clean industrialization. For her, the solution proposed by the report is palliative and does not adequately affect the causes of the problem.

When considering the decision on the responsibilities, strategies and methods to achieve the sustainable development, the debate branches out into three basic positions. A statesmanlike view, which considers environmental quality a public good that should be standardized, regulated and promoted

---

[118] HERCULANO, 1992.

by the State; community vision, in which civil society organisations must play a predominant role in the transition to a sustainable society based on the nation that there be sustainable development without democracy and social participation. And a market view that states that market mechanisms and relationships between producers and consumers are the most efficient means to drive and regulate the sustainability of development.[119]

Despite the criticisms of sustainable development over the years, it is currently crucial to seek practical actions and moral values that associate economic development with planetary sustainability. Sustainability will continue to be an urgent social and economic challenge for the coming decades.

Here, we need to criticize past economic development based on policies that governments have unfairly/ unwisely granted to oil industries. This has been a major factor contributing to our global climate crisis.

---

[119] VIOLA & LEIS, 1995.

According to 2018 UN report about the world governments' responsibilities to reduce global emissions forty percent in the next ten years, the reality has been showed us another scenario. There has been an increase of global CO_2 emissions around the world. And in an analysis by Hallam (2019) on the climate he concluded that since 90s, the world's present political systems have facilitated a sixty percent increase in global CO_2 emissions.

Our greatest challenges ahead are concentrated in the transformation of philosophy and discourse into action and realization with real time metrics and benchmarks and penalties. It is in this logic that obstacles and great disagreements are now found, justice, environmental sustainability, economic viability, participatory democracy, behavioral ethics, solidarity and integrative knowledge.

Our aspirations depend on the level and quality of public awareness, our perceptions of reality and problems, and our capacity for organization to drive changes towards a sustainable society. It will also depend on the ability of social movements to attract forces, to establish alliances and to lead a process that makes the philosophy of sustainability a real alternative to pure economic development.

And finally, negativism and resistance to science has been a path taken by many governments and societies around the world. This must stop.

Science negativism it is based on an unconscious denial of truth about climate crisis that our planet is facing. All adults can feel the radical environmental change through natural consequences that we have lived through. Hurricanes, flooding, wildfire, melting of the poles, rising sea levels, and temperature increase are the most common events that all of us have witnessed.

Recycling of course is very important, but we need more effective actions. Governments need to take responsibility for all of us. They must assign and achieve environmental goals to each country. Recycling and preservation are not enough. We must change the energy matrix to clean energies, invest in sustainable development politics for all public measures, and eradicate poverty.

If we want to build a better world for future generations, we need to do this now. Now is the time not just in for real and urgent actions. Together, we can do this. Together we are stronger. Together we must do this.

6

THE NEW ENVIRONMENTAL ERA

NEW RULES TO FLEEING FROM CHAOS: SUSTAINABILITY PARTNERS

6.1 Environmental education, recycling, and popular participation

The environmental issue is still little known broadly by the world population and basically reaches the only most privileged classes of society. The contemporary problem of climate change is increasingly highlighting the theme of the environment and challenges for the construction of a consistent environmental education.[120]

Environmental education is a comprehensive form of education, which aims to reach all citizens, through a permanent participatory pedagogical process that seeks to instill a critical

---

[120] LOUREIRO; TORRES, 2014.

awareness about the environmental problems, materials present on overview of the genesis and evolution of today's environmental problems.

Few people know, but Environmental Education has been part of the curriculum in Brazil for a long time. Law 9,795 of 04/27/1999 established the National Environmental Education Policy, which states that all levels of education and the community in general are entitled to environmental education and that the media must collaborate in the dissemination of this information.

The preservation of the environment depends on everyone: government, educators, companies, Non-Governmental Organizations (NGOs), media and every citizen. Environmental education is fundamental in solving these problems, as it will encourage citizens to know and do their part, including: avoiding waste of water, light and unnecessary consumption, selective collection, purchasing products from companies concerned about the environment, charging the competent authorities to enforce the law, treating garbage and sewage correctly, protecting natural areas , making land use plans, and encouraging recycling, among others.

Environmental education emphasizes local norms, and seeks to maintain respect for the different ecosystems and human cultures of Earth. The duty to recognize global similarities, while effectively interacting with local specifics, is summarized in the following motto: Think globally, act locally.

An environmental education program to be effective must simultaneously promote the development of knowledge, attitudes and skills necessary for the preservation and improvement of environmental quality. We start small, with the school laboratory, an urban metabolism and its creating a natural and physical resources. We expand through the vicinity and successively to the city, region, country, continent and planet.

Learning is always more effective if the activity is adapted to the real life situations of the particular environment in which students and teachers live.

In 1975, as an offshoot of the Stockholm Conference, the International Seminar on Environmental Education took place in Belgrade, where the conceptual bases of environmental education were defined. Guimarães (1995) made the following statement:

> The basic principle of environmental education is attention to the natural and artificial environment, considering ecological, political, social, cultural and aesthetic factors. Environmental education should be continuous, multidisciplinary, integrated within regional differences, focused on national interests and focused on questioning the type of development. Its priority goal is the formation in individuals of a collective consciousness, capable of discerning the environmental importance in the preservation of the human species and, above all, stimulating cooperative behavior in different inter- and intra-nation relations.

Some concepts used even exhaustively in Environmental Education are not always understood in their full meaning and generate serious conceptual problems of reasoning formation. Following are some of them:

It is necessary that the process of environmental education is carried out with real, active participation of the students. Brugger (1994) defines as training does not happen: a type of instruction where people are led to perform a certain type of function or task, identified with a certain utilitarian-one-dimensional pattern of thought-action. It is ineffective as aspirations and objectives, which by their content transcend the established universe of word and action, are reduced to terms of that universe.

To carry out an education process, whether environmental or not, it is necessary to create a concept of general interest that is strengthened incrementally. As the dimensions of education for active citizenship are absorbed the effect will be the multiplication of participation in the decisive processes of public interest.[121]

Thomas Jefferson was a strong proponent of national public education. He advocated providing a formal education as a basis for lifelong learning, a pursuit he believed represented humanity's purest endeavor.[122]

---

[121] JACOBI, 1999.
[122] SAYLAN; BLUMSTEIN, 2011.

In this view, the process of participation is only realized through the knowledge of its cause, and access to information. Only by inclusiveness, especially from more excluded social groups, can we promote the behavioral changes necessary to enable more action oriented by the general public. Well-informed citizens, by assuming themselves to be relevant actors, are more able to pressure authorities, as well as to motivate themselves for actions of co-responsibility and community participation.[123]

It is necessary to organize and formulate strategies based on participatory processes, between governmental organizations and NGOs. This is the only way to maintain socio-political commitment aimed at achieving environmental management objectives; include to the most deprived strata of the population.

Based on this process, Jacobi (1999) explains that the need to strengthen the institutional context is unquestionable. Jacobi posits that for this

---

[123] JACOBI, 1999.

to occur it is essential to generate references for residents regarding the availability, access and costs of services locally. This will allow them to establish various links with the perception of environmental problems in their most immediate surroundings - neighborhood and home.

The existing challenge is to formulate viable links for the implementation of improvements that are both technical and socially efficient. Institutional strengthening is a relevant but not sufficient condition to consolidate improvements. There is a need to take into account the level of information and/or misinformation of residents about interrelations between the environment and their involvement and motivations with a perspective that emphasizes the general interest.[124]

Institutional reforms and, fundamentally, new methods in the processes of management of environmental issues will be important for the success of environmental policies.[125] Thinking about this

---

[124] JACOBI, 1999.

[125] CAHN, 1995.

possibility, there is an urgent need for the adoption of educational measures that are directed to the betterment of the environment. We must not forget humans have been exercising decisions affecting the environment for millennia and evaluating the impact has not always been harmonic.

Paradigm shifts require changes in attitude.[126] Environmental education seeks to implement new relationships between men and other species, men with the abiotic environment, and most importantly man and other men.

Environmental education must start at home, reach the street and the block, encompass the neighborhood, cover the city, overcome the peripheries, rethink the fate of pockets of poverty, penetrate the intimacy of oppressive spaces, reach the peculiarities and regional diversities to integrate national spaces. The basic aim of environmental education is to ensure a healthy environment for all men and types of life existing on the face of the Earth.

---

[126] LOUREIRO ET AL, 2002.

Freire (2002) develops educational praxis through dialectics. This is a process of reflection and action in the construction of knowledge; it is fundamental way to intervene in reality and promote change leading to citizenship. This implies back-and-forth movement between the critical unfolding of reality and the transforming social action. All this is based on the principle that human beings educate themselves reciprocally and mediated by the observable world.

Loureiro et al (2002) state that environmental education is an educational and social praxis that aims to construct values, concepts, skill and attitudes that enable the understanding of the reality of life, plus the lucid and responsible action of individual and collective social actors on the environment. In this sense, it contributes to the attempt to implement a civilizational and societal pattern distinct from the current one, based on a new ethics of the relationship between society and nature. Thus, for the real transformation of the structural and conjunctural matrix in which we live, environmental education, by definition, is a strategic element in the formation of broad critical awareness of social and production relations that balance human needs and nature.

For education, including environmental education, to be effective, it cannot be something verticalized in the educator-educating model. But rather it must be reflective, constructed:

> "From this process, comes a knowledge that is critical, because it was obtained in an authentically reflective way, and implies a constant act of unveiling reality, positioning itself in it. Knowledge constructed in this way perceives the need to transform the world, because in this way men discover themselves as historical beings."[127]

Environmental education understood as learning new skills that lead to changing values and attitudes should involve all who interact with the environment: the production sector, government, and organized civil society.

As a social movement, the environmental issue consolidated in the late 1960s, and in the early

---

[127] FREIRE, 1983.

1970s. But it is since only the 1980s that it was popularized.

Certainly, the intensification of socio-environmental problems has impacted global public opinion and attracted attention to a reality little observed until then. It is a fact that the multiplication of socio-environmental problems has contributed to the growth of ecological awareness. Simultaneously we observe an increase in questions about the relationship between society and nature and about the disintegration of knowledge about economics, ecology, sociology and biology, in the sense of the approximation of the natural and social sciences.

However, this awakening of ecological awareness has not yet resulted in significant changes in the direction of government policies and individual lifestyles.

Nevertheless, ecological awareness incrementally grows and materializes in social movements, scientific initiatives, the media, government policies, international organizations and business activities.

Environmentalism is analyzed by Viola e Leis (1991), who note that the movement began, in the 1970s, with minorities of scholars and environmentalists. This small group organized itself around the denunciation of aggressions and the defense of ecosystems. Eventually a critical mass of scholars coalesced to question the capitalist conquest of new spaces, until acquiring a true multisectoral cohert. From the point of view of the concerns and guiding themes of the movement, the focus of attention broadened to include issues such as political ecology, demographics, the relationship between poverty and ecology, technical-scientific issues, ethics, north-south relations, and the search for a new model of development.

The multisectoral profile does not imply uniformity of positions, it indicates only a growing plurality of social sectors that recognize the legitimacy of the field and the need to include it in the planning of national and global development.[128]

---

[128] VIOLA E LEIS, 1991.

Guimarães (1995), points out that it is necessary to exercise praxis in environmental education, because only action generates activism. Study alone generates an immobility that will not fulfill the transformative possibility of education. Thus, the solution would be to participate in a true dialogue between the reflexive attitude and the action of theory with practice, that is, thinking with doing. This process strengthens human knowledge and enables us to alter our destiny.

The following sentence by Freire (1983) seems very appropriate to the concept exposed:

> Man is a being of relationships. Culture is a reflection of man's creative process and this creative process makes him an agent of active adaptation and not of an accommodation. This conception distinguishes nature from culture, understanding culture as the result of its work, of its creative effort. This discovery is responsible for the rescue of self-esteem, because, both culture is the work of a great sculptor, as the brick made by the potter. We try to overcome the dichotomy between theory and practice, because during the process, when man discovers that his practice supposes knowledge, he concludes that knowing is interfering in reality, he perceives himself as a subject of history.

Environmental learning can occur inside or outside of school, through curricula about also social movements, including the current activism.

When we fly the flags on environmental causes we trumpet the truth as way produce attitude change. This is not a simple thing to do.

People around the world concerned with climate change, do not do anything except recycle. Although recycling is very important from the perspective of environmental preservation, we need to go beyond that. It is necessary to reduce consumption and actively participate in political debates on environmental issues. So, consciously, we can collaborate with the preservation of life, the recovery of environmental and salvation of the Earth.

We must promote a great movement regarding environmental concerns. If governments don't listen, then the people should act. We believe that only environmental education, environmental management and environmental movements through widespread activism will contribute to the improvement of the quality of our world. Dialogic discourse leads to towards an educational group participation which leads to practice, i.e., a new relationship with nature from a dialectical society. This is a point that we can do in an organized goals, only through joint actions.

Greta Thunberg has said in some of her speeches about climate change: "We must admit that we do not have this situation under control. We must admit that we are losing this battle. We must stop playing with words and numbers, because we no longer have time for that."[129] "I felt like I was the only who cared about the climate and the ecological crisis."[130] "Change is coming whether you like it or not."[131] "Yet you all come to us young people for hope."[132]

Greta is Swedish. She is just a teenager, but her actions have impacted all the world through her words and actions. She asks hard questions about environmental causes and consequences that need

---

[129] HALLAM, 2019.
[130] BBC, 2018.
[131] UN, 2019.
[132] UN, 2019.

change. She challenges leaders at all societies and governments. She is 17 years old and her voice became a clarion call to millions of other teens, adults, people around the world, like you and me.

Greta's actions went viral after her protests in front of the Swedish parliament asking her government to commit to carbon emissions targets agreed to in 2015, in Paris. She held a sign saying: "School Strike for Climate". She started skipping school on Fridays and asked for the support of all students all over the world. In 2019, she sailed on yacht for two weeks to New York to participate in the UN Climate conference. She made a scathing speech to the UN demanding answers from world leaders.

The best pedagogy involves action-reflection-action. Thus, start at home, reach the street and the block, encompass the neighborhood, cover the city, overcome the peripheries, rethink the fate of pockets of poverty, penetrate the intimacy of oppressive spaces, reach the peculiarities and regional diversities to integrate the national spaces of education. Observe, participate, analyze, reflect. Repeat.

We know environmental movements are extremely important as instruments for the dissemination of environmental education. We know collective awareness practical actions, discourse and participation provide the path we must travel to preservation, maintenance and recovery of nature. We know dynamic individuals like Greta play an important role.

Certainly, environmental activists, non-governmental organizations, environmental educators and governments all really need to do their part in protecting society, defending life of the planet and fighting climate emergencies. Most governments historically favored industrial development and were slow to notice and protect the natural world.

Our eyes are now wide open to see that our civilization need more attention, more action, and more opposition to political rhetoric about protecting the economy while ignoring climate emergency.

Climate emergency is not a rhetorical call for accelerated climate action. It's a call for a major transition of the economy.[133] And, as in all complex dynamic systems, this is not an "either-or" choice.

Right now we need to eliminate use of fossil fuel replace it with renewable energy. Right now we need to step up and to recover biodiversity, protect water sources, plants and micro-ecosystems, and forests promoting life of species. Right now we need to invest dramatically in sustainable development public and private. Right now we need to replace the energy matrix and eliminate greenhouse gases.

Together in only one voice, you and I, including the whole society, companies, schools, universities, environmental and social activists, we must maintain momentum with popular actions and movements in favor of preservation and protection of the Earth. Right now we need to listen to the science and hear the cry of nature against its devastation.

---

[133] HALLAM, 2019.

We have the power: our writed voice! This is the great strength of to democracy. Together we're stronger! The planet needs each one of us, right now!

6.2 Science: knowing the truth!

For many years The Economist magazine has been a global source of discussion of politics, arts, science, law, diplomacy and technology. Founded in 1843 by James Wilson, a hatter from the small Scottish town of Hawick it was an early proponent free trade, internationalism and minimal political and economic interference by the state. A striking aspect of The Economist is its irreverence for example, using cartoons to explain its ideas.

The scientist and astronomer Joseph Norman Lockyer created Nature magazine in 1869. Nature consolidates different research articles from fields such physics, chemistry, geology and biology. Its main goals was to provide information about advances made in any area of natural knowledge as well as an opportunity to discuss political and ethical implications of scientific discoveries.[134]

By the end of the 19th century, interest in scientific issues began winning space from the society pages of print media.

In July of 1880 the magazine Science was created by journalist John Michels. He based it on Nature, the magazine from Europe that had been making great strides in the publishing environment.[135] This magazine was incorporated by the American Association for the Advancement of Science (AAAS) in 1901. According to AAAS (2013) the magazine became an important factor making the AAAS a world recognized entity during the entire 20th century.

Nature, Science, and The Economist began publishing research on climate data in the 1970s.

From 1970 until the present date environmental research and analysis has been presented in these periodicals. The data collected by

---

[134] BARATA, 2010.
[135] BARATA, 2010.

Colacios (2014) in his research on climate change in both Nature and Science was prominent from 1970 to 2005.

With the rise in awareness through these publications the world began global attempts to address these issues in international meetings, conferences, and summits. Solution were sought to avoid the environmental impacts caused by accelerated industrial development and other human factor negatively affecting the planet.

Also starting in the 1970s world opinion on the oil and gas industry, its pollution, and severe consequences for humanity, ecosystems and planet's life with burning fossil fuel there was the diffusion of new sources of energy started shifting.

These alternative technologies aimed to reduce the dependence of modern society on the consumption of fossil fuels through use of renewable energy sources.

Nature and Science have been important sources of climate change research for 80 years. These magazines have provided a steady flow of information and scientific proof on climate science.

The data presented below by Colacios (2014) from Nature and Scientific three subthemes: general, ozone layer, and global warming from 1970 to 2005. It shows the number of articles about climate crisis by category.

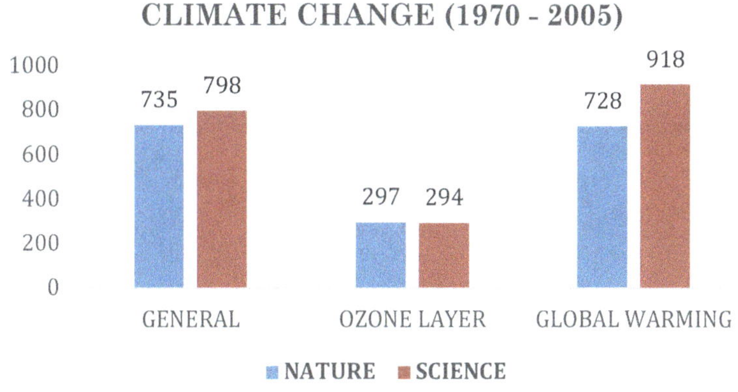

16 Climate Change by Nature and Science Magazines (1970-2005)

Source: COLACIOS, 2014. Created by author, 2021.

Climate change is the most urgent, most serious and deepest dimension of the environmental crisis of the 21st century. It is the greatest threat to our modern world, and it has been concerning for all people. Temperature increases will cause irreversible changes on ecosystems worldwide. The climate crisis is more serious than anticipated and it is threatening natural ecosystems and the very fate of humanity.[136]

The global temperature has increased, in the last century, by 0.74°C, and is expected to increase by 4.3 ± 0.7°C in the next 100 years.[137] The consequences of nature modified through concentrated greenhouse gases at nonacceptable levels in the atmosphere, marine, and terrestrial environment can be irreversible. These consequences can also lead to a catastrophe, far beyond the possible control of humans.[138]

In the figure 16 we show that both magazines, Nature and Science, produced technical-scientific information committed to the worldwide discussion about the environment at the same period, from 1970 to 2005. Between the sub themes discussed about Climate Change - General, Ozone Layer, and Global Warming, Nature has produced fewer scientific articles than Science. The total number in Nature

---

[136] IPCC 2019.
[137] PACIFICI et al., 2015.
[138] STEFFEN et al. 2018.

magazine was 1,760 research articles, and in Science, 2,010 articles. This difference was mainly represented by the global warming theme, where Nature produced 728 and Science, 918 scientific works.

In terms of adoption of international measures allied to the climate science the 1970s were an important legal and scientific landmark. The Stockholm Conference in 1972 boosted the scale of international environmental debates. Other discussions around the world have increased over time too. Now, in 2021, the climate emergency has become the main cause of scientific alarm.

According to Colacios's (2014) sources, Nature and Science have published many scientific articles about climate change analyzed in figures 17 and 18. To understand these graphs more fully, we must integrate some important world events that occurred.

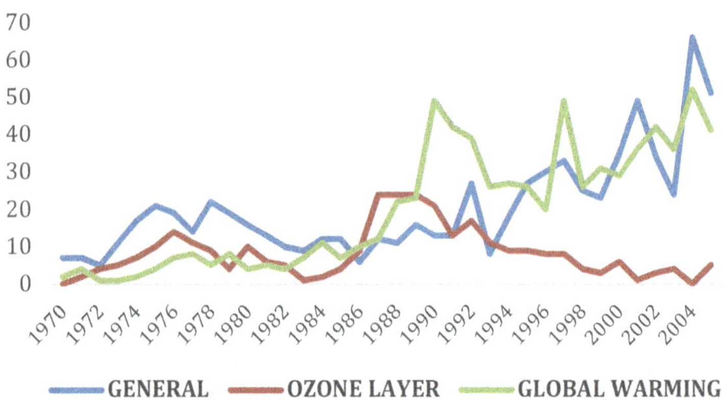

17 Climate Change by Nature Magazine

Source: COLACIOS, 2014. Created by author, 2021.

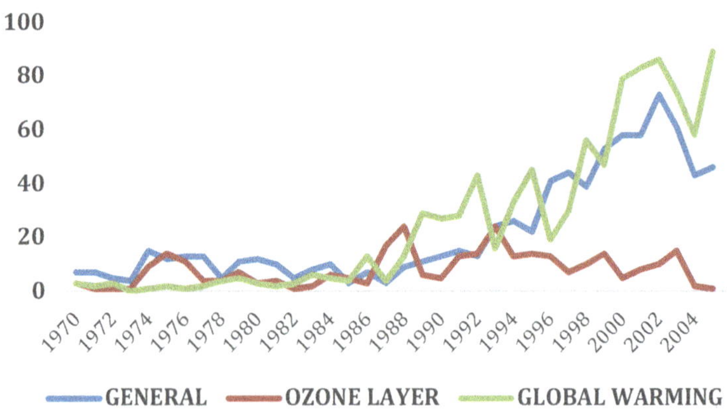

18 Climate Change by Science Magazine

Source: COLACIOS, 2014. Created by author, 2021.

By the 1980s, emissions of greenhouse gases had caused concern among scientists who were relating them to climate change resulting from anthropic activities.

The Montreal Protocol, signed in 1987 in London, and ratified in 1989 with several revisions, is an international treaty whose objective is to protect the ozone layer through a decrease in polluting gases. Chlorofluorocarbons (CFCs) are one of the most destructive gases. Currently 197 countries have ratified the Montreal Protocol.

In the figure 17 by Nature Magazine, the Ozone Layer was the theme of 297 research articles over 35 years. In the period from 1987 to 1997, that is, 10 years after the Montreal Protocol, Ozone Layer research reached 54 percent of all articles. This demonstrates that scientific publications reflect public opinion, often provoking debate, legislation, and international agreements.

Following the same line of reasoning figure 18 reveals that Science produced 294 papers on the Ozone Layer theme during at that same period, 1970-2005. Then, 10 years after the Montreal Protocol, its research Science also published an increased percentage of Ozone Layer themed papers, with 51 percent. This striking correlation reaffirms that scientific research is the best way to educate the general public and create legislative momentum.

In 1988 the IPCC[139] aiming to supply scientific information about climate was created. The IPCC scientists provided much-needed scientific support on annual climate change, its impacts, and unintended consequences. With data importantly, the IPCC proposed methods to mitigate damage and to reduce risks.

The Intergovernmental Panel on Climate Change (IPCC) highlighted how anthropical activities have influenced the global warming through greenhouse gas emission (GHC), mid-twentieth century. According to the IPCC, the increase in the CO_2 concentration comes from emissions by the burning fossil fuel by man. Human activity accounts for three quarters of total CO_2 emissions.

---

[139] The IPCC is an organization of governments that are members of the United Nations or WMO. The Intergovernmental Panel on Climate Change (IPCC) is the United Nations body for assessing the science related to climate change.

In 1992 the United Nations Framework Convention on Climate Change (UNFCCC) was born. Due necessity to actions regulation by executive power was implanted the Conference of the Parties which is an association of all member countries signatories to the Convention, or Convention supreme organ, its aimed to make more energetic decisions on climate.

According to the United Nations Framework Convention on Climate Change (UNFCCC):

> Changes in climate, directly or indirectly, are attributed to the activity of humans that alter the composition of the global atmosphere and that override the natural variability observable in equivalent time periods.

In figure 17, by Nature magazine research from 1990 to 2005, it shows we see that global warming themed works totaled 88 percent of all from 1970-2005. We understand this number represents the growth of concern about global warming in the science community. As a result, legislation, international agreements and meetings also that were stepping up at this time.

In Science magazine (figure 18), the research from 1990 to 2005 shows that its global warming articles totaled 78 percent of the whole period from 1970 to 2005. It is reasonable to concluded that global warming has been a growing concern in science communities, governments, students, and people all around the world.

It is appropriate that Science magazine began addressing biotic ramifications of global warming, including the possibility of reducing the atmospheric load of carbon dioxide by forest management.[140]

And finally, the analysis on Climate Change with the General theme of the magazines Nature (figure 17) and Science (figure 18) shows that the peak of scientific articles occurred after the 2000s. Nature magazine had its general themed research record from 2004 to 2005, with a total of 16 percent of the 35 year span. Science magazine reached its

---

[140] COLACIOS, 2014.

maximum production in the years 2002 to 2003, totaling 17 percent.

According to Colacios (2014) this growth in the amount of research over time implies an increase in the interest of readers and, more broadly, throughout society, over for years.

In fact, for Colacios (2014), his research on Nature and Science magazines is this:

> What is interesting with these quantitative data is to show that there was a significant presence of research, news and opinions on climate change on the pages of the Nature magazine and that there was a conformity between the data of both magazines (Nature and Science) showing that the cut-off historical period was significant for this type of concern among scientists, journalists and politicians; this is evident when observing that there has been a significant increase in texts over the thirty-five years analyzed.[141]

In short, it is clear that environmental issues have gained more space and importance in global matters in recent years. Climate related events, agreements and meetings have clearly been based on solid science accumulated over the years.

---

[141] COLACIOS, 2014.

The truth we need to accept has been explained by scientists for years. Climate change is a real concern worldwide. The climate emergency is now deep and worsening. Unfortunately, greenhouse gases emissions have been increasing more quickly.

Scientists began to investigate global climate change on a micro level only from 1990s. This is when they developed greater technological methods to stud trends and possible solutions.[142]

It is important to say there are still some controversies when the discussion comes to climate change. Some people believe climate change is ongoing. There is another group of people who don't believe in it at all. However, the climate crisis is a reality that we all need to acknowledge, understand, and combat.

The article published in 2010 in the Proceedings of the National Academy of Science says that 97 to 98 percent of all scientific papers on climate science are in accord with the Intergovernmental

---

[142] ADEGER, 2001.

Panel on Climate Change (IPCC); furthermore, these papers overwhelming agree that these climate modifications that have been recorded have been a strongly influenced by human activities.[143]

We have to believe in the science! We must understand this climate crisis is not under control!

The fact that people literally feel the effects of atmospheric weather conditions on their skin, makes everyone have a false impression of intimacy with this issue.[144] Climate change is not equivalent to daily weather.

According to the Intergovernmental Panel on Climate Change (IPCC) to avoid incalculable suffering due to climate change, it is urgent to conserve our biosphere through a huge increase of scale in efforts. We have not a moment to waste!

In 2020 scientists presented at BioScience some measures which could help us to understand and contain the crisis, such as:

---

[143] USP, 2019.
[144] Nunes, 2003.

1- Establishing high taxes to discouraging the consumption of fossil fuel, replacing it with renewable energy;

2- Legally limiting the emission of short-term pollutants, such as methane, soot and hydrofluorocarbons;

3- Preserving and restoring Earth's ecosystems such as phytoplankton, coral reefs, forests, savannas, mangroves and swamps that absorb CO_2.

4- Consuming more food based on fruits, vegetables, grains and oilseeds and less on animal protein.

5- Containing the extensive extraction of raw materials and the overexploitation of ecosystems as sources of economic growth in favor of the long-term maintenance of our biosphere.

6- Establishing family planning policies to stabilize population growth.

Although discussions around environmental issues have been spreading through conferences all over the world, so far we all have failed in taking sufficient actions. Scientists also say they are disappointed by this reality but do see some cause for optimism. According to the signatory scientists from all around the world they have been encouraged by a recent global wave of concern about the climate, such as, governments adopting new policies, children carrying out strikes, lawsuits moving forward, and citizen movements demanding change.[145]

Among the numerous technical-scientific works and discussions on global climate, no effective action has been taken to date to combat the environmental chaos that is a result of the current capitalist world growth models.

In the analyzes exposed in this book, where we raise historical, scientific, legal, cultural, political, ideological and social issues, we must conclude that the problems of the environment are reflections of a narrow economic, political and industrial perspective of contemporary society.

In summary according to the analyzes developed throughout this book, we are on a path has brought about negative consequences to humanity,

---

[145] BIOSCIENSE, 2020.

the environment, and the planet. In order to discuss and raise awareness about the need to change attitudes and understand the chaotic process that the planet has been going through over the years, we note here that the biggest challenge we face at this moment is global warming. We believe that together we will be able to find a successful solution and that it is as urgent as it is necessary.

Finally, the silence of skeptics and the murmur of those who oppose science on climate will not help to resolve the environmental crisis. The ideal does not exist, but the possible is the essential path to the salvation of the planet. This depends on all of us, particularly on governments, research, the science community and its funding, industries and all the people that inhabit the wonderful planet Earth.

ABOUT THE AUTHOR

Professor doctor Barbara Lobo is an educator and environmental scientist from Brazil. She has been living in the United States since 2016.

Bárbara Lobo was born in Bom Jesus do Itabapoana/ Rio de Janeiro, Brazil; she is the mother of Luiza and Gabriel (Tyler's father) and wife of Alberto Lobo. Her family is her foundation. She is a daughter and granddaughter of writers; therefore she grew up in a literary world where she discovered a taste for literature.

Doctor Lobo has a doctoral degree (Doctoral - PhD from the Americana University and Post-Doctoral from the Tres de Febrero University) in Educational Sciences and a master's degree in Environment and Health Sciences from the Plinio Leite University Center.

She taught for fifteen years at Estacio de Sa University in Brazil teaching for undergraduate and postgraduate courses for Environmental Sciences, Environmental Engineering, Environmental Health, among others.

Currently, she is doing original research on environmental science and has written several books on higher education, environmental science, and others, available on Amazon.

REFERENCES

ALMINO, J. Naturezas Mortas: A filosofia política do ecologismo. Brasília: Fundação Alexandre Gusmão, 1993.

ALVES, Rodrigo Couto. A (in)viabilidade de consórcios públicos intermunicipais para a gestão de resíduos sólidos no amazonas. Dissertação apresentada ao Programa de Pós-Graduação - Mestrado em Ciências do Ambiente e Sustentabilidade na Amazônia. Universidade Federal do Amazonas: Manaus, 2020.

ANDRADE, Pedro Gustavo Gomes. O Meio Ambiente Nos Tribunais Internacionais: Diálogo de Jurisdições e Unidade do Sistema Jurídico. Dissertação apresentada ao Programa de Pós-Graduação Stricto Sensu da Faculdade de Direito da Universidade Federal de Minas Gerais. Belo Horizonte, 2017.

BARBIERI, J. C. Desenvolvimento e meio ambiente. Petrópolis: Vozes, 1997.

BBC, Brazil News. Os quinze países que mais emitiram CO2 nos últimos 20 anos. (Disponível em: https://www.bbc.com/portuguese/brasil-50811386.amp . Acesso em 2021)

BELCHIOR, Germana Parente Neiva; MATIAS, João Luis Nogueira. O princípio da solidariedade como marco jurídico-constitucional do Estado de direito ambiental. In: HAUSCHILD, Mauro Luciano; GUEDES, Jefferson Carús; RODRIGUES JÚNIOR, Otavio Luiz. (Org.). Meio ambiente, propriedade e agronegócio. 1 ed. Brasília: IP Editora, 2011, v. 1.

BIOSCIENCE. 2020. World Scientists' Warming of a Climate Emergency. (https://academic.oup.com/bioscience).

BLUMS. The Tension Between Liberalism and Environmental Policymaking in the United States. New York: State University of New York, 1995.

BNDES. Análise das diversas Tecnologias de tratamento e disposição final de resíduos sólidos urbanos no Brasil, Europa, Estados Unidos e Japão. Jaboatão dos Guararapes, PE: Grupo de Resíduos Sólidos - UFPE, 2014.

BOFF, Leonardo. Ecologia: grito da terra, grito dos pobres. São Paulo: Ática, 1999.

______________. Saber cuidar. Ética do humano: compaixão pela terra. Petrópolis: Vozes, 2000.

BRASIL, Paula Emília Moura Aragão De Sousa. A Proteção do Meio Ambiente Como Dever Fundamental. Tese da Universidade Federal Do Ceará Apresentada Ao Programa De Pós Paula Emília Moura Aragão De Sousa Brasil. Fortaleza. 2016

BRÚSEKE, Franz Josef. Desestruturação e Desenvolvimento. In: VIOLA, E. & FERREIRA, L. C. (Orgs.) Incertezas de Sustentabilidade na Globalização. Campinas: Unicamp, 1996.

BRUGGER, Paula. Educação ou adestramento ambiental. Coleção teses. Letras contemporâneas. Ilha de Santa Catarina: 1994. 141p.

BUARQUE, Cristovam. A desordem do progresso: o fim da era dos economistas e a construção do futuro. Rio de Janeiro: Paz e Terra, 1990.

CAHN, M. A. The Tension Between Liberalism and Environmental Policymaking in the United States. New York: State University of New York, 1995.

CAPRA, Fritjof. O ponto de mutação. São Paulo: Cultrix, 1982.

CARDOSO, Saulo Barbosa. Desenvolvimento Sustentável: Variáveis e Caminhos a Longo Prazo. TCC apresentado ao Curso de Aperfeiçoamento para Oficiais de Máquinas do Centro de Instrução Almirante Graça Aranha. Marinha do Brasil: Rio de Janeiro, 2016.

CAULLEY, D. N. Document Analysis in Program Evaluation. Portland: Or. Northwest Regional Educational Laboratory, 1981.

CAVALCANTI, Clóvis. Política de Governo para o desenvolvimento sustentável: Uma introdução ao tema e a esta obra coletiva. São Paulo: Cortez, 2002.

CLEVELAND, C. J. Capital humano, capital natural e limites biofísicos no processo econômico. In: CAVALCANTI, Clóvis. Política de Governo para o desenvolvimento sustentável: Uma introdução ao tema e a esta obra coletiva. São Paulo: Cortez, 2002.

COLACIOS, R. D. Um clima de incertezas: as controvérsias científicas sobre mudanças climáticas nas revistas Science e Nature (1970-2005). Tese de Doutorado em História Social. São Paulo: USP - Universidade de São Paulo, 2014.

CONSTITUIÇÃO DA REPÚBLICA PORTUGUESA. (Disponível em http://www.ministeriopublico.pt/iframe/constituicao-da-republica-portuguesa, acesso em 2020).

CONSTITUIÇÃO FEDERAL DO BRASIL. (Disponível em: www.jusbrasil.com.br. Acesso em 2020).

CORREIA DE ANDRADE, Manoel. O desafio ecológico: utopia e realidade. São Paulo: HUCITEC, 1994.

COZETTI, N., Lixo- marca incomoda de modernidade, Revista Ecologia e Desenvolvimento, 96: 2001.

CRUZ, G. D., As riquezas que jogamos fora, Revista Ecologia e Desenvolvimento, 77, 46-51, 2001.

DE SOUZA, Ricardo Timm. Totalidade e desagregação: sobre as fronteiras do pensamento e suas alternativas. Porto Alegre: EDIPUCRS, 1996.

DENNINGER, Erhard. - Security, diversity, solidarity instead of freedom, equality, fraternity‖, in Constellattions, volume 7, número 4, Oxford: Blackwell Publishers ltd, 2000, P. 507/521, tradução de Christopher Long e William E. Schuerman.

DIAS, G. F. Educação Ambiental: princípios e práticas. São Paulo: Gaia, 1992.

__________ Pegada Ecológica e sustentabilidade humana. Editora Gaia. São Paulo- SP. 257p. 2002.

FIORILLO, Celso Antonio Pacheco. Curso de Direito Ambiental Brasileiro. 5ª edição. São Paulo: Saraiva, 2004.

FISHER, Elizabeth. Environmental Law: A very short introduction. New York: Oxford University Press, 2017.

FORANTTINI, O. P., Ecologia, epidemiologia e sociedade. São Paulo: Edusp, 1992.

FOLADORI, G., Um Olhar Antropológico sobre a questão ambiental, MANA,323-348, 2004, acessado http://www.scielo.br/

FREIRE, Paulo. Educação e Mudança. Rio de Janeiro: Paz e Terra, 1983. 11ed.

__________________Pedagogia do Oprimido. 34ª edição. São Paulo: Paz e Terra, 2002.

FURTADO, Celso. O mito do desenvolvimento econômico. Rio de Janeiro: Paz e Terra, 1996.

GIL, Antônio Carlos. Como Elaborar Projetos de Pesquisa. São Paulo: Atlas, 1991.

GONÇALVES, Carlos Walter Porto. Natureza e sociedade: elementos para uma ética da sustentabilidade. In: COIMBRA, José de Ávila Aguiar (Org.) Fronteiras da Ética. São Paulo: Senac, 2002.

______________________________ Os (des) caminhos do meio ambiente, São Paulo, 2ª edição, Contextos, 1990.

GUBA, E. G. e LINCOLN, Y. S. Effective Evaluation. San Francisco, Ca: Jossey-Bass, 1981.

GUDYNAS, Eduardo. Ética, ambiente e ecologia: uma crise entrelaçada. Revista Eclesiástica Brasileira. Petrópolis: Vozes, n° 52, Fascículo 205 – Mar, 1992, p. 68-69.

GUIMARÃES, Mauro. A Dimensão Ambiental Na Educação. Campinas, Sp: Papirus, 1995 (Coleção Magistério: formação e trabalho pedagógico. 1995. 107p).

GUIMARÃES, Roberto P. Assimetria dos interesses compartilhados: América Latina e a agenda global do meio ambiente. In: LEIS, H. R. (Org.). Ecologia e política mundial. Rio de Janeiro: Vozes/FASE, 1991.

HERCULANO, S. C. Do desenvolvimento (in) suportável à sociedade feliz. In: GOLDENBERG, M. (Org.). Ecologia, ciência e política. Rio de Janeiro: Revan, 1992.

IBGE. Estimativas de População dos municípios para 2018. (Disponível em https://agenciadenoticias.ibge.gov.br/agencia-sala-de-imprensa/2013-agencia-de-noticias/releases/22374-ibge-divulga-as-estimativas-de-populacao-dos-municipios-para-2018).

[IPCC] Intergovernmental Panel on Climate Change. 2019. Climate Change and Land IPCC. Google Scholar.

______________________________. Climate Change. (Available in: ipcc.ch. Acessed in: jan. 2021).

IRVING, M. A. Projeto Sana Sustentável: Uma iniciativa de base comunitária. In: D'ÁVILA, M. & PEDRO, R. Tecendo o desenvolvimento: Saberes, gênero e ecologia social. Coleção Eicos. Rio de Janeiro: Bapera, 2003.

JACOBI, P. Cidade e Meio Ambiente: percepções e práticas em São Paulo. São Paulo: Annablume,1999.

JONAS, Hans. El principio de responsabilidad: Ensayo de uma ética para la civilización tecnológica. Barcelona: Herder, 1995.

LA TORRE, M. Antonietta. Ecología y moral. La irrupción de la instancia ecológica en la ética de Occidente. Bilbao: Editorial Desclée de Brouwer, 1993.

LAGO, Antônio & PÁDUA, José Augusto. O que é ecologia. São Paulo: Brasiliense, 1992.

LEIS AMBIENTAIS BRASILEIRAS. (Disponível em http://www.estrategiaods.org.br/as-sete-principais-leis-ambientais-brasileiras/; https://iusnatura.com.br/principais-leis-ambientais/. Acesso em 2020).

LIMA, M. F. C. Desenvolvimento Sustentável, a crise do fordismo e os países periféricos. In: RODRIGUES, A. M. (Org.). Meio ambiente ecos da ECO. Campinas: Unicamp, 1993.

LOUREIRO, C. F. B; TORRES, J. R. Educação Ambiental dialogando com Paulo Freire. Cortez Editora, 2014.

LÜDKE, Menga e ANDRÉ, Marli. Pesquisa em Educação: abordagens qualitativas. São Paulo: E.P.U., 1986.

MACHADO, C.J.S., Sociedade, Meio Ambiente e Políticas Públicas e a erradicação das velhas práticas, Jornal da Ciência, 2003, acessado em http://www.jornaldaciencia.org.br.

MACHADO, Paulo A. Leme. Direito ambiental brasileiro. São Paulo: Malheiros, 2002, pp. 127-128.

MARRIOTT, Betty B. Environmental impact assessment: a practical guide. United States, 1997.

MATIAS, João Luis Nogueira. Em busca de uma sociedade livre, justa e solidaria: a função ambiental como forma de conciliação entre o direito de propriedade e o direito ao meio ambiente sadio. In: NOGUERIA, João Luís (coord.); SALES, Tainah Simões; AGUIAR, Ana Cecília Bezerra de (org.).

Ordem econômica na perspectiva dos direitos fundamentais. Curitiba, PR: CRV, 2013.

MATIAS, J. L. N.; MATIAS, J. F.N. A convergência entre os direitos de propriedade e ao meio ambiente sadio: a cessão de uso das águas da União para a produção de pescado no Brasil. In: A Efetivação do direito de propriedade para o desenvolvimento sustentável: relatos e proposições. Florianópolis: Fundação Boiteux, 2010.

MATIAS, João Luís Nogueira; MATTEI, Júlia. Aspectos comparativos da proteção ambiental no Brasil e na Alemanha. Nomos. Revista do Programa de pós-graduação em Direito da UFC, v. 34, n.2, jul./dez. 2014, p. 227/244.

MAY, Tim. Pesquisa Social: questões, métodos e processos. 3. ed. Porto Alegre: Atrmed, 2004.

MEIRELES, S., Crimes ambientais, os ganhos dos acordos judiciais, Revista Ecologia e Desenvolvimento, 92: 2001.

MIRANDA, E.E., DORADO A. J., ASSUMPÇÃO J.V., Doenças Respiratórias crônicas em quatro municípios paulistas. Campinas: Ecoforça – 1994.

MINISTERIO DO MEIO AMBIENTE, www.ministeriomeioambiente.gov.br , acessado em agosto de 2005.

MOREIRA, A C. M. L., Conceitos De Ambiente E De Impacto Ambiental Aplicáveis Ao Meio Urbano, 2004, acessado em http://www.usp.br.

MORIM, Edgar & KERN, Brigitte. Terra – Pátria. Porto Alegre: Sulina, 1995.

MORIN, Edgar & TERRENA, Marcos. Saberes globais e saberes locais. 3ª edição. Tradução de Paula Yone Stroh. Rio de Janeiro: Garamond, 2001.

MUCCI, J. L. N., Introdução às Ciências Ambientais, São Paulo, Manole, 2005.

NEIVA, A, MOREIRA, M., COZETTI, N., MEIRELLES, S., NORONHA, S., Mineiro, P., Agenda 21, o futuro que o brasileiro quer, Revista Ecologia e Desenvolvimento, 93: 2001.

OLIVEIRA, Júlio Cesar de Melo. SOFT SENSOR VEICULAR PARA EMISSÃO DE MEDIÇÕES DE GASES. Dissertação de Mestrado Engenharia Elétrica e de Computação. Universidade Federal Rio Grande do Norte: Natal, 2017.

O'NEILL, Kate. Waste. UK: Cambridge, 2019.

PACIFICI M, Foden WB, Visconti P, Watson JEM, Butchart SHM et al. Assessing species vulnerability to climate change. Nature Climate Change, 2015.

PECCEI, Aurélio & IKEDA, Daisaku. Antes que seja tarde demais. Rio de Janeiro: Record, 1984.

PELICIOONI, A F. O movimento ambientalista e a educação ambiental, São Paulo, Manole, 2005.

PHILIPPI JR. (ORG) Saneamento do meio. São Paulo: Fundacentro, 1988.

PHILIPPI JR. & MALHEIROS, T. F., saúde Ambiental e Desenvolvimento. São Paulo: USP, 2005.

POPULAÇÃO MUNDIAL. Relatório da ONU em 2019. (Disponível em: https://nacoesunidas.org/populacao-mundial-deve-chegar-a-97-bilhoes-de-pessoas-em-2050-diz-relatorio-da-onu/amp/#)

ROEGEN, Geosgescu. The entropy law and the economic process. Cambridge: Havard University, 1971.

SACHS, Ignacy. Ecodesenvolvimento: crescer sem destruir. São Paulo: Vértice, 1986.

SACHS, Jefrey D. The age of sustainable development. New York: Columbia University Press, 2015.

SANTOS, Alexandre Rosa. Climatologia. 2005.

SAYLAN, C; BLUMSTEIN, D. T. The failure of environmental education. University of California Press, 2011.

SOBRAL, HR, SILVA CCA. Balanço sobre a situação do meio ambiente na metrópole de São Paulo. Projeto de Tese de Doutorado, Departamento de Geografia: UFRJ, 2001. São Paulo, Perspec, 1989.

STAHEL, A. W. Capitalismo e entropia: os aspectos ideológicos de uma contradição e uma busca de alternativas sustentáveis. In: CAVALCANTI, Clóvis (Org.). Desenvolvimento e natureza: estudos para uma sociedade sustentável. São Paulo: Cortez, 1995.

STEFFEN, W. et al. Trajectories of the Earth System in the Anthropocene. Proceedings of the National Academy of Sciences. 2018.

TOYNBEE, Arnold. A sociedade do futuro. 2ª edição. Rio de Janeiro: Zahar Editores, 1974.

UNEP. Consumo crescente em escala regional. (Disponível em: www.mma.gov.br/. Acesso em: jan 2020).

VIEIRA, P. F. Meio ambiente, desenvolvimento e planejamento. In: Meio ambiente, desenvolvimento e cidadania: desafios para as ciências sociais. São Paulo: Cortez, 1995.

VIOLA, Eduardo & LEIS, Hector. A evolução das políticas ambientais no Brasil, 1971 – 1991: do bissetorialismo preservacionista para o multissetorialismo orientado para o desenvolvimento sustentável. In: HOGAN, D. & VIEIRA, P. F. (Orgs). Dilemas socioambientais e desenvolvimento sustentável. Campinas: Unicamp, 1995.

__________________________. Desordem global da biosfera e a nova ordem internacional: o papel organizador do ecologismo. In: LEIS, H. R. (Org.). Ecologia e política mundial. Rio de Janeiro: Vozes/FASE, 1991.

VIOLA, Eduardo. Reflexões sobre os dilemas do Brasil na segunda metade da década de 1990 e sobre uma agenda de políticas públicas baseadas na democracia, na eqüidade, na eficiência e na sustentabilidade. Anais. Meio Ambiente, desenvolvimento e política de governo – Workshop. Olinda: Fundação Joaquim Nabuco, 1996.

VITOR, C. A questão ambiental deve estar no centro de tudo, Revista Ecologia e Desenvolvimento, 100: 2002.

WORLD BANK. CO2 emissions (kt). (Available in: databank.worldbank.org. Acessed in: jan. 2021).

______________. Total greenhouse emissions (kt of CO2 equivalent). (Available in: databank.worldbank.org. Acessed in: jan. 2021).

______________. What a waste 2.0 – A global snapshot of solid waste management to 2050. (Available in: openknowledge.world.org/handle/10986/2174. Acessed in oct 2020).

www.ingramcontent.com/pod-product-compliance
Lightning Source LLC
Chambersburg PA
CBHW070800160726

48004CB00001B/267